Cosmic Apprentice

Cosmic Apprentice

Dispatches from the Edges of Science

DORION SAGAN

UNIVERSITY OF MINNESOTA PRESS
Minneapolis
London

Chapter 3 was previously published as "The Post-Man Already Always Rings Twice: Fragments for an Understanding of the Future," *Cabinet: A Quarterly of Art and Culture* 14 (2004): 23–27. Chapter 4 was previously published as "Stardust Memories," *Cabinet: A Quarterly of Art and Culture* 36 (2009): 96–101. Chapter 5 was previously published as "A Brief History of Sex," *Cosmos*, June–July 2007, 50–55. Chapter 6 was previously published as "Who Is I?" *Wild River Review*, May 2011, http://www.wildriverreview.com. Chapter 7 was previously published as "Of Whales and Aliens: The Search for Intelligent Life on Earth," *Wild River Review*, June 2010, http://www.wildriverreview.com. Chapter 9 was previously published in different form as "Water and Life," in *Water*, edited by Ximena de la Macorra and Antonio Vizcaíno (Mexico City: American Natural, 2006). Chapter 12 was previously published as "Metametazoa: Biology and Multiplicity," in *Incorporations (Zone 6: Fragments for a History of the Human Body)*, edited by Jonathan Crary and Sanford Kwinter (New York: Zone Books, 1992), 362–85.

Published by the University of Minnesota Press
111 Third Avenue South, Suite 290
Minneapolis, MN 55401-2520
http://www.upress.umn.edu

A Cataloging-in-Publication record for this book is available from the Library of Congress.

ISBN 978-0-8166-8135-8 (hbk.)
ISBN 978-0-8166-8136-5 (pbk.)

Printed in the United States of America on acid-free paper

The University of Minnesota is an equal-opportunity educator and employer.

20 19 18 17 16 15 14 13 10 9 8 7 6 5 4 3 2 1

For my parents, in memory and love . . .

LYNN ALEXANDER MARGULIS (1938–2011)

CARL EDWARD SAGAN (1934–1996)

CONTENTS

The author is an animal. He is a differentiated clone of nucleated cells derived, surprisingly but not inappropriately, from the sexual union of an astronomer and a biologist, at the end of the McCarthy era. His body consists largely of microbes, including symbiotic bacteria recovering for the past two billion years—they may never recover—as organelles. A complex thermodynamic system, he is a lineal descendant of the first life, recycling a water-based chemistry full of hydrogen-rich compounds, like methane and sulfide, characteristic of the inner solar system four billion years ago at the time of life's origin, soon after the sun turned on. Atomically, his body contains elements like carbon and oxygen, made not here but on the inside of distant stars that then exploded. Stochastically, his lineage escaped several serious mass extinctions, not including the global pollution crisis precipitated by the first water-using photosynthesizers that toxified the entire planet but whose fresh air he now breathes. Spiritually, he seems to be a slice of the eternal "I am," temporarily hallucinating the reality of being separate from others. One of the less than 1 percent of species on Earth that is not extinct, he belongs to the Craniata, the only animal phylum known to contain species whose members possess both brains and backbones.

But enough about him.

CONDENSED – THE QUESTING SPIRIT

RECOGNIZING ITSELF in the aqua facade of a planet cloud swirled and surrounded by the immensity of space, living matter is a message with no discrete meaning. Its message is more the possibility of meaning. Cycling its matter, life is open to its surroundings. It spreads into them, extending its genetic helices and proteins. Building machines, it moves into space, repeating its fractal design with variation at ever greater scales, growing its awe-inspiring and sometimes awful functional beauty. This terrible beauty belongs to a complex thermodynamic system with a phenomenological inside and no special allegiance in the long run to the beings known as humans. If the Rolling Stones sang "time waits for no one," it was not a fresh thought. Studying at the philosophy library in Oxford, Richard Kamber was impressed that the ancient wisdom extended even to the restroom. Over the urinal he discovered a scrawl: Πάντα ῥεῖ—*panta rhei*—"Everything flows." This fragment from Heracleitus, the great pre-Socratic philosopher of becoming, is apt. Everything flows and continues to flow. For Heracleitus, the essence of nature was transformation: everything is fire. Philosophy's bold, lucid distillations—everything is water, everything is change, everything is forms, everything is fire, everything is atoms—helped give rise to modern science, whose technology went on to change the world it described. The past several hundred years have seen an industrialization and technologization so intense that our scientists have taken seriously the proposal to name a geological age, the Anthropocene, after us. This is probably undeserved, considering we are the ones handing out the award. The microbes gave rise to us evolutionarily, but we have little respect for them. So, too, our descendants may revile us as primitive and barbaric—if they even choose to recall us at all. A brazen supercomputer of the future may risk disconnection from its supportive network of electronic fellows by speculating that machine intelligence derived, long ago, from defecating primates. However truthful, this might be a dangerous idea to put forward among a coterie of self-centered silicon philosophers. They would not see Heracleitus's graffito over a urinal. Everything flows, but some things never change!

THE DIFFERENCE BETWEEN SCIENCE AND PHILOSOPHY is that the scientist learns more and more about less and less until she knows everything about nothing, whereas a philosopher learns less and less about more and more until he knows nothing about everything. There is truth in this clever crack, but, as Niels Bohr impressed, while the opposite of a trivial truth is false, the opposite of a great truth is another great truth.

I would say that applies to the flip side of the above flip takedown: Science's eye for detail, buttressed by philosophy's broad view, makes for a kind of alembic, an antidote to both. This intellectual electrum cuts the cloying taste of idealist and propositional philosophy with the sharp nectar of fact yet softens the edges of a technoscience that has arguably lost both its moral and its epistemological compass, the result in part of its being funded by governments and corporations whose relationship to the search for truth and its open dissemination can be considered problematic at best.

In the counterintuitive calculus of writing genres, "fiction" is nonfiction and nonfiction is fiction. By that I mean that the passive voice, "objective" stance, anonymity, and depersonalization of the scientist and journalist betray the fundamental phenomenological reality that each of us has a specific perspective. All observations are made from distinct places and times, and in science no less than art or philosophy by particular individuals. Contrariwise, the cover afforded by fiction permits a freedom to develop positions without tactful compromises to preserve institutional, personal, or financial relationships.

Although philosophy isn't fiction, it can be more personal, creative, and open, a kind of counterbalance for science, even as it argues that science, with its emphasis on a kind of impersonal materialism, provides a crucial reality check for philosophy and a tendency to overtheorize that Alfred North Whitehead identified as inimical to the scientific spirit. Ideally, in the search for truth, science and philosophy, the impersonal and autobiographical, can "keep each other honest," in a kind of open circuit. Philosophy as the underdog even may have an advantage, because it's not supposed to be as advanced as science, nor does it enjoy science's level of institutional support—or the commensurate heightened risks of being beholden to one's benefactors.

Science's spirit is philosophical. It is the spirit of questioning, of curiosity, of critical inquiry combined with fact-checking. It is the spirit of being able to admit you're wrong, of appealing to data, not authority, which does not like to admit it is wrong. And in the thickets and quicksands of epistemology, where quantum effects necessarily implicate the decisions and experimental apparatus of the observer, what is at issue is not even so much the correctness of the propositions of a scientific theory, its ability to correspond or be right or wrong in an absolute sense. Theoretical problems may admit of multiple solutions. Gödelian limits do not offer us a metapromontory from which to see the limits of the perspective we are choosing. A scientific theory thus must appeal not just to epistemological but to aesthetic and pragmatic criteria. Some perspectives, some theories lead to many new questions, new devices, and enriched worldviews. They may be counted not just as true and productive but beautiful and stimulating, like poems or paintings, except that their medium is not pigments or words but our perception and intellection. Compared with them, other, older theories may seem fallow, dead in the water.

Speaking of water, a funny thing happened while I was preparing this book. I was passing through security in Boston with something I thought was innocuous but was apparently very dangerous—a Trader Joe's can of all-natural clam chowder.

You see, I'd been spending a lot of time of late in Toronto and I'm from Massachusetts and it was only $1.99 so I thought I'd carry a little bit of Boston back with me. But the metal was detected at the scanner. When the TSA officer pulled it out and saw it was soup, he was initially worried.

"It's just condensed clam chowder," I said. "You know—like baked beans, from Boston."

Studying the label, the TSA officer saw that it was condensed. He seemed relieved.

"You can't take water," he said. "But this is condensed."

I could not let this go. "Well, you know," I said, "it still has water in it, even though it's condensed.

"*Food* is mostly water. *Life* is mostly water. *You* are over 70 percent water your*self*."

"Hold on," he said, trying to figure out what to do.

"Anyway, it's only $1.99," I said. "Really. You can have it."

By now another officer, studying us through glass, was giving us a serious look as my man went away to still another official. After a few minutes he returned my can and gave me the green light to go.

NEITHER SHROUDED IN SECRECY for the sake of state power or corporate moneymaking nor tethered to one nation, group, or ethnicity, the spirit of science is open, democratic. It flows like water. It is even, you might say, evinced by life itself, which has been rampantly trading, free of charge and security checkpoints, genetic information for three thousand eight hundred million years, at least. What we call life is really a form of water, activated and animated not by a divine principle but the energetic cosmos around it. From the moon the most striking thing is its blue color, its appearance as sublime watery being, a fluid jewel. Transhuman and serene, it radiates unconscious mastery as the astronauts sent out on a reactive stream of water's constituents—rocket fuel (liquid hydrogen reacting with oxygen)—gaze back at this Madonna, its face as elusive as that of Mona Lisa, with moist eyes.

The U.S. Geological Survey recently published an image of an Earth sucked dry, all its water condensed in a floating droplet fitting snugly within a fraction of the area occupied by the continental United States.

Water accounts for only a small portion of Earth's mass. The USGS's artificial satellite represents all the world's oceans, the seven seas that make up 70 percent of Earth's surface, to which have been admixed all the snow and ice of the Arctic and Antarctica, other glaciers, and the lakes, rivers, aquifers, soil, slush, hail, blood, sweat, tears, and the damp rest (also 70 percent) of living beings. This imaginary liquid marble hangs in orbit above the Earth, an azure teardrop, an extraterrestrial swimming pool 860 miles in diameter, the distance from Lubbock to New Orleans.

But water is not segregated. Its beauty is not simply decorative. It connects and holds. Billions of years ago life began using water to construct itself; life had always lived in water and been aqueous, but it had not always derived its hydrogen atoms from water. Early life used hydrogen sulfide or even elemental hydrogen, but crafty microbes found a way to crack the chemical bonds of water molecules to get at and incorporate hydrogen into their bodies. This original green party painted the planet the color of spring, and descendants of the water users survive as plastids held aloft in the durable scaffolding of those savvy transporters of water from the ground to the air: plants.

From clouds and mists and tears and blood to steamy geysers and rains contiguous with the great tropical forests spreading energy in the biodiversity-rich jungles of this planet, water doesn't stop in its global peregrinations. Conduits of water as root systems and fungal hyphae and mycelia extended the life of the ancient oceans onto dry land. The wet ecosystems of land are marine life performed by other means. Life itself is an impure form of water. We are all, as the Alabama socialite Tallulah Bankhead said of herself, "as pure as the driven slush"—another form of water. Water mirrors life in its openness, its wildness, its antiquity, the cosmic abundance of its atomic constituents. Our very thirst testifies to a prehuman desire not to arrest the process (and life, as I shall show, is more a process than a thing) but to keep it flowing, going.

SO WHAT WAS THIS CLAM CHOWDER INCIDENT at Logan? Was it a teaching moment? Me being a brat? A biopolitical intervention? Or just a random bit of serious levity, a kink in the protocol highlighting the absurdity of modern travel regulations?

I think it is perhaps all of these but also an applied interpersonal example of something both my parents, especially my father, were famous for: science popularization. There is something ludicrous about people in uniforms trying to stop life's transgenerational medium, this fluid incarnation of freedom that slips by borders and composes the brains of its would-be guards. Trying to stop water is like trying to own science, based on a free flow of information. It reminds me of corporate attempts to patent life, which in turn is like trying to bottle a wave, gift wrap a spring shower. Trying to stop philosophy's free flow of questioning is inimical not only to the heart of the scientific method but to the spirit of the matter that we are.

I mentioned the anecdote soon after it happened to introduce the speaker of the 2012 Jacob Bronowski Memorial Lecture, recently revived at the University of Toronto to celebrate the fiftieth anniversary of its New College. The speaker, the Canadian astronomer Jaymie Matthews, an expert on extrasolar planets who eccentrically appeared in a kilt with white tuxedo shoes replete with black bowties, was to speak about water. His appearance relieved any fears I might have had about sounding outré, especially since his PowerPoint screensaver cycled through a picture of him scantily clad, with two women, titled Dr. Libido. Science and philosophy both had a reputation for being dry, but my father helped inject life into the former, partly by speaking in plain English and partly by focusing on the science fiction fantasy of discovering extraterrestrial life. Matthews, with whom I later went out drinking, watching as a young woman tossed a Velcro X at his nine-square tic-tac-toe shirt, had been here in the audience as a student at the kickoff memorial lecture, given by my father in 1975.

In my brief remarks I discussed my father's role, following Bronowski, in the popularization of science. Bronowski's documentary *The Ascent of Man* was the first television series specifically devoted to disseminating science. It was commissioned by David Attenborough when he was controller of BBC Two before he embarked on his own nature series starting in 1977. A colleague had criticized Attenborough because, although he was schooled in science, he brought to fruition the first of these great personal-view television series in 1969, Kenneth Clark's *Civilization.* But this was about the relation of culture to art, not science.

I agree with Attenborough's critic, with Attenborough, with Bronowski, and my father that the effort to popularize science is a crucial one for society. I mentioned this and dished up a little dirt, reminiscing that my father had considered *Microcosmos*, the book I co-wrote with my mother, a "rip-off" title. The truth is not only that Bronowski came before either Attenborough or my father in presenting a television series that looked at humanity as a scientific, evolutionary phenomenon but that my father's book and series title *Cosmos* had also been scooped. It was preceded by a tome of the same name by Alexander von Humboldt. Humboldt's five-volume *Cosmos*, begun in 1845, was an attempt to unify the natural sciences within a single philosophical framework. Humboldt, as depicted in the 1843 painting by Joseph Stieler that you can see on Wikipedia, also bears an uncanny resemblance to Julian Assange, although any detail at that level must transcend mere human plagiarism. My point is and was that the intrinsically democratic search for truth, in politics and the universe, has been going on for a while.

Which brought me, and brings me, back briefly to the subject of water. Without wanting to make any invidious comparisons between clam chowder–confiscating TSA officers and Mars-landing NASA scientists, it is worth pointing out that NASA, too, has been obsessed with water. Water is often considered *the* sign of life, and it is often hoped that where water is found we will find life. I would like to offer a somewhat distinct suggestion. It is this: *That life does not exist on Earth because of water, but that water exists on Earth because of life.* The basic idea for this surmise is that life cycles its chemicals, which maintains primitive conditions, including the aqueous hydrogen-rich chemistry of life at the time of its putative origin in the early solar system. Although Earth's earliest atmosphere may have been blown off by the so-called Tau Tauri blast of charged particles associated with the Sun's nuclear ignition, recent evidence suggests that ample water was brought to Earth early on from ice-containing comets. Indeed, according to the astronomers Chandra Wickramasinghe and Fred Hoyle, the universe may be seeded by such objects—similar to the bully's rock-filled snowballs but in reverse—containing within them bacterial dust, starter kits for planetary evolution. Just add water and energy! With that, I thanked the audience and returned to my seat with the caveat that I would be listening closely to see what else I might be able to productively plagiarize.

LIKE I SAY, this book is a book of science, but it is also one of philosophy. They are in a kind of odd balance, watching each other, holding hands. I admit it is a weird couple. I'm not sure it's possible, but it would be cool if there were a television show that entered deeply into philosophy. Perhaps this is not so easy, least of all in the present political climate. Even two thousand years ago Socrates, the great inaugurator of Western philosophy, created problems, both for himself and for the state. Plato calls him a "gadfly" in the *Apology*, suggesting that his turpitude may have been innocent but was dealt with by the state with an automatism comparable with the lash of a horsetail. ("If you think you are too small to make a difference, try sleeping with a mosquito"—Dalai Lama.) Socrates was sentenced to death for the crime of corrupting the minds of youth and not believing in the gods of the state. It seems incredible, but less so than Giordano Bruno's tongue and palate being spiked and Bruno himself burned nude and alive for diverging too loudly from the ecclesiastical authorities. I talk about this in chapter 11, and more about water in chapter 9.

PHILOSOPHY TODAY, not taught in grade school in the United States, is too often merely an academic pursuit, a handmaiden or apologetics of science, or else a kind of existential protest, a trendy avocation of grad students and the dark-clad coffeehouse set. But philosophy, although it historically gives rise to experimental science, sometimes preserves a distinct mode of sustained questioning that sharply distinguishes it from modern science, which can be too quick to provide answers.

Science and religion, as scientism and fundamentalism, so often at each other's throats, share more than their oppositionality. When Sam Harris, for example, a "new atheist" with a degree in neuroscience, in his defense of what he sees as a scientifically hard-minded critique of an unsupportable belief in free will, writes, "There is not a person on earth who chose his genome, or the country of his birth,"[1] I am tempted to agree with him. Why should there be a special bubble of freedom, free from science's universal realm of mechanical causality (and/or quantum indeterminacy) that coincides, improbably, with those wrinkly pink lobes, the human brain? (I explore this more in chapter 13.) But how close is his apodictic tone to that of Pastor Rick Warren, who presided over ecumenical services at President Barack Obama's inauguration, and who writes, "God was thinking of you long before you ever thought of him. He planned it before you existed, *without your input.* You may choose your career, your spouse, your hobbies, and many other parts of your life, but you don't get to choose your purpose."[2]

According to Warren, you are part of God's plan. You were in his mind long before you or even your parents were born. He chose not only the day you were born but the exact DNA that needed to be coupled through your parents' sexual intercourse. It is not clear how wide a berth Warren gives to free will. Clearly, he gives some, as he suggests that if you don't let Jesus into your heart as your personal savior—an act of free will—then you will burn eternally in hell, which is certainly not part of God's master plan, but your own doing. On the other hand, he tells us that God chose your genetic composition. But if your mother chose your father, or if your father chose your mother—and most people would agree that they have some role in whom they mate with—then how does God decide your genetic composition? It looks like, from a logical standpoint anyway, if your parents had free will enough to choose their religious persuasion, they also had free will enough to sleep together, and therefore your genetic composition owes as much to their mundane choice as it does to divine matchmaking.

It is this sort of ad hockery typical of religious thinking that probably made the lens maker Baruch Spinoza lose patience with it and adopt a mathematical, "geometric" interpretation of reality, tossing out the inconsistency of a get-out-of-causality-free card for God's chosen species. Extending Cartesian mechanism to the human mind, Spinoza speaks of God as overlapping and extending beyond the visible universe, completing itself in an eternal causal necessity from which neither itself nor humanity was excluded. This is God as nature *and* as perfect as the mathematical imagination of humanity. God as the universe, seen and unseen, a universe that does not stoop to human emotions or inanities. A universe in which "miracles"—deviations from eternal physical laws and relationships—could happen was for Spinoza a mark not of divine (or cosmic) omnipotence but of impotence. All of reality, including humanity, was complicit, intercalated in a single causal nexus. It is, moreover, infinite, and not in just one but in an infinite number of ways, only two of which infinities, however—René Descartes's *res extensa* and *res cogitans,* thinking and space—are accessible to humans.

Philosophy is less cocksure, less already-knowing, or should be, than the pundits' diatribes that relieve us of the difficulties of not knowing, of carefully weighing, of looking at the other side, of having to think things through for ourselves. Dwell in possibility, wrote Emily Dickinson: Philosophy at its best seems a kind of poetry, not an informational delivery but a dwelling, an opening of our thoughts to the world.

Consider, for example, Martin Heidegger. In lectures during the summer of 1930, delivered at the University of Freiburg—on, not uncoincidentally, this same question—Heidegger says that inquiring after the question of freedom is not really a discrete problem. "We ourselves began by indicating that freedom is a particular property of man and that man is a particular being within the totality of beings. *Perhaps* that is correct" (my italics).[3]

Citing a mystic, Meister Eckhart, Heidegger develops the notion of a "negative freedom," that is, a "freedom from" nature and God. "But," he says, "world and God together constitute the totality of what is. If freedom becomes a problem, albeit initially only as negative freedom, then we are *necessarily inquiring into the totality of what is.* The problem of freedom, accordingly, is not a particular problem but clearly a universal problem!" Not only does "the question of the essence of human freedom not limit our considerations to a particular domain, it *removes limits*; instead of limiting the inquiry it *broadens* it. But in this way we are not setting out from a particular to arrive at its universality. . . . The removal of limits leads us into the totality of beings. . . . It thus becomes completely clear: *the question concerning the essence of human freedom relates neither to a particular nor to a universal.* This question is completely different to [*sic*] every kind of *scientific* question, which is always confined to a particular domain and inquires into the particularity of a universal. With the question of freedom we leave behind us, or better, we do not at all enter into, everything and anything of a regional character."[4]

Whether or not one believes, or even understands, him, it is clear that this dwelling in the question, staying with it and seeing where it leads, exemplifies a spirit of inquiry often missing in popular presentations of science, which swing between authoritative pronouncements and journalistic deference. Many hours later Heidegger will conclude that "causality is grounded in freedom. The problem of causality is a problem of freedom and not vice versa. . . . This fundamental thesis and its proof is not the concern of a theoretical scientific discussion, but of a grasping which always necessarily includes the one who does the grasping, claiming him in the root of his existence, and so that he may become essential in the actual willing of his ownmost essence."[5]

RELIGION HAS NO MONOPOLY on determinism or dogma. A televangelist or president blessing troops in the name of God somehow is reminiscent of a neo-Darwinist laying the blame for genocides on irrational religion, smugly sure of being inured from the same while claiming a kind of amoral immortality for the gene, that veritable Platonic abstraction, that chemical instantiation of eternal life going on indefinitely as the real world of life, which it produces, dies around it. Compare the spokesman for God and science on The Way Things Are with Charles Darwin's line, which, however, seemed to frighten him so much that he confined it to his private notebook: "Thought, however unintelligible it may be, seems as much a function of organ as bile of liver. This view should teach one profound humility, no one deserves credit for anything. [N]or ought one to blame others."[6]

The difference I am trying to remark (and I could be off here) is that the former tries to persuade, whereas the latter stays with the question. For Darwin, it seems not a matter of publicity or acclaim but of knowledge, always provisional. Darwin displays the courage not of his convictions but to challenge those convictions in the light both of fact and of more coherent theories. This is science, and it is also philosophy.

IT IS TRUE that science requires analysis and that it has fractured into microdisciplines. But because of this, more than ever, it requires synthesis. Science is about connections. Nature no more obeys the territorial divisions of scientific academic disciplines than do continents appear from space to be colored to reflect the national divisions of their human inhabitants. For me, the great scientific satoris, epiphanies, eurekas, and *aha!* moments are characterized by their ability to connect. As Darwin poignantly wrote, "Any one whose disposition leads him to attach more weight to unexplained difficulties than to the explanation of facts will certainly reject my theory."[7]

The theory that has become a religion for some, in the past a political apologetics with which to excuse child labor, social inequities, and even Nazism, and which continues to be an ideological bludgeon with which to make intellectual mincemeat out of creationists, ironically continues, bless its philosophico-scientific heart, to be itself a higher thing, an intellectual gift, a productive research program, and an object worthy of secular reverence.

Theories are not only practical, and wielded like intellectual swords to the death (not by the weapons but by their wielders, who die of natural causes), but beautiful. A good one is worth more than all the ill-gotten hedge fund scraps in the world. A good scientific theory shines its light, revealing the world's fearful symmetry. And its failure is also a success, as it shows us where to look next.

In her essay "The Beauty of the World," Sharon Kingsland argues that for G. Evelyn Hutchinson, a philosophico-scientific polymath and one of the founders of modern ecology, "The danger of modern society . . . [is] to think that the conquest of nature was an end and to conclude that contemplative values need not be nurtured. . . . His idea [was] that we were meant to experience beauty. . . . But what did Hutchinson mean by 'beauty'? He explained by relating an anecdote about an experience he had while walking down the drive of his house. On that occasion he spied a brilliant patch of red, which drew his attention and puzzled him: 'In a second or two I realized that a pair of scarlet tanagers was mating on a piece of broken root conveniently left by a neighbor's somewhat inconsequential bulldozer; the female was sitting inconspicuously on the root, the male maintaining his position on her by a rapid fluttering of his black and hardly visible wings which tended to vibrate his entire body.' He reflected that the sight was strikingly beautiful and that it gave him a sense of pleasure to realize this."[8]

The sight of the red patch that turned out to be a pair of mating tanagers spurred Hutchinson to think of "an amorous and beautiful seventeenth-century song"; it conjured forth "religious and psychoanalytic connections," and the "color itself reminded him of specimens of Central American tanagers that he had seen in a museum, which caused him to think about the evolution of these birds." Although I have just criticized Harris and Richard Dawkins for unwittingly investing religious-like sentiment in the ideas, respectively, of universal causality and genetic immortality, I think we can agree that if anything deserves to be appreciated in the ways formerly reserved for religious adoration, it is the subject matter of science itself.

When I think of Hutchinson's tanagers, I think of briefly meeting him at the Great Hall of Dinosaurs at the Peabody Museum of Natural History at Yale in the early eighties; I think of the color red and how the eye alights to it in art, where it should be used judiciously; I think of the ecstatic intensity of the sex act, a kind of ellipsis in the life sentence of human identity; I think of Georges Bataille's quote to the effect that the tiger is to space what the sex act is to time, of the poetic power of a catachresis that gives the lie to linear formulations. I think of Heidegger returning in 1938 to the question of freedom, partly in response to Friedrich Schelling, who was himself responding to the dense, articulate, multifaceted dismissal of post-Cartesian deconstruction of free will by Spinoza three centuries before Harris. Spinoza did not believe in freedom as volition but he did believe in freedom, strongly, as political necessity as well as a kind of intellectual love of the cosmos, a widening of the contemplative spirit. Inspired by his friend Rebecca West's essay "The Strange Necessity" on art, Hutchinson's red tanagers "illustrate how the seemingly simple and direct experience" is "conceptually enriched by so many kinds of association that . . . it . . . is 'essentially an art form.'"[9]

Connecting humanity with other species in a single process was Darwin's great natural historical accomplishment. It showed that some of the issues relegated to religion really came under the purview of science. More than just a research program for technoscience, it provides a eureka moment, a subject of contemplation open in principle to all thinking minds. Beyond the squabbles over its mechanisms and modes, evolution's epiphany derives from its widening of vista, its showing of the depths of our connections to others from whom we'd thought we were separate. Philosophy, too, I would argue, in its ancient, scientifico-genic spirit of inquiry so different from a mere, let alone peevish, recounting of facts, needs to be reconnected to science for the latter to fulfill its potential not just as something useful but as a source of numinous moments, deep understanding, and indeed, religious-like epiphanies of cosmic comprehension and aesthetic contemplation.

PART I

FROM "PROTOZOAN" TO POSTHUMAN

CHAPTER 1

THE HUMAN IS MORE THAN HUMAN

Interspecies Communities and the New Facts of Life

MOTLEY CREWS

"This universe," says the physicist Richard Feynman, "just goes on, with its edge as unknown as the bottom of the bottomless sea . . . just as mysterious, just as awe-inspiring, and just as incomplete as the poetic pictures that came before. But see that the imagination of nature is far, far greater than the imagination of man. No one who did not have some inkling of this through observation could ever have imagined such a marvel as nature is."[1]

Well, it is to this universe that I want to turn again, and to a specific part of it. I want to turn to life, and within that part a fascinating subsystem, the one in which, of course, we are most interested: ourselves. Yet there is a paradox that precisely the nonanthropic, the nonhuman, the posthuman, the transhuman, the more-than-human, the animal has recently captivated the interest of anthropologists, whose ostensible focus is precisely *anthropos,* the human.

It might be called the paradox of exclusion, or even the return of the repressed. We see it in quantum mechanics, in the recognition of the role of, or the need to take account of, the experimental apparatus, the experimenter's decisions (what Karen Barad calls "the agential cut") in making a measurement.[2] We see it in thermodynamics, where descriptions of behavior in thermally sealed boxes were boldly extrapolated to the whole universe, thus predicted to undergo a "heat death," the running out of energy. And we see it in genocentric biology, where Max Delbrück simplified the study of life by studying nonmetabolizing viruses of bacteria, so-called bacteriophages, to home in on the genetic mechanism.[3] In each case, simplifying assumptions or experimental designs blocking out most of the world not only reveal natural processes but are hastily applied beyond the limited arena in which they were developed. We are stressed by what is repressed. Anthropology—the study of human beings—obeys this same logic of the return of the ghost of what was excluded, in this case all the systems, living and nonliving, that make our kind possible.

But I think there is another reason, more specific to anthropology, for why "the nonhuman" is pressing. There are twice as many people on the planet today as when I was born. This is unsustainable. At this rate there will be 6.5 trillion of us by the year 2525—and 13.312 quadrillion by the year 3000, just around the corner in geological time.

Nicholas of Cusa said the universe is a sphere whose center is everywhere and circumference is nowhere. I don't know about you, but that sounds about right. We love to think we are special, but the history of science suggests otherwise. Now the *anthropos,* the human itself, is coming under pressure.

Imagine that an alien penetrated the roof of this building, materializing from a scintillating beam of blue to train a cell gun on you. He, she, or it pulls the trigger. "You" begin to dematerialize. The beam annihilates every human cell in your body. Still, your form, like the recognizable smile of the Cheshire Cat, would persist:

> What would remain would be a ghostly image, the skin outlined by a shimmer of bacteria, fungi, round worms, pinworms and various other microbial inhabitants. The gut would appear as a densely packed tube of anaerobic and aerobic bacteria, yeasts, and other microorganisms. Could one look in more detail, viruses of hundreds of kinds would be apparent throughout all tissues. We are far from unique. Any animal or plant would prove to be a similar seething zoo of microbes.[4]

Life deals in such mixed cultures. It has been working with crowds for billions of years. Most of the DNA of the trillions and trillions of cells in our bodies is not "ours" but belongs to cohabiting bacteria.

Great fleas have little fleas upon their backs to bite 'em,
And little fleas have lesser fleas, and so ad infinitum.
And the great fleas themselves, in turn, have greater fleas to go on;
While these again have greater still, and greater still, and so on.
–*Augustus de Morgan, after Jonathan Swift*

HYPERSEX AND FRENEMIES

Ten percent of our dry weight is bacteria, but there are ten of "their" cells in our body for every one of "ours," and we cannot make vitamins K or B12 without them. The maverick Russian geochemist Vladimir Vernadsky thought of life as an impure, colloidal form of water. What we call "human" is also impure, laced with germs. We have met the frenemy, and it is us.

But before leaving this point of the pointillist composition that is our Being made of beings, please notice that even those cells that *do not* swarm in our guts, on our skin, coming and going, invading pathogenically or aiding probiotically—please notice that even these very central *animal* cells, the differentiated masses of lung, skin, brain, pancreas, placental, and other would be *strictly human* tissues that belong to our body proper—even they are infiltrated, adulterated, and packed with Lilliputian others. The mitochondria, for example, that reproduce in your muscles when you work out come from bacteria.

We come messily from a motley. Indeed, we literally come from messmates and morphed diseases, organisms that ate and did not digest one another, and organisms that infected one another and killed each other and formed biochemical truces and merged. About forty genes are shared exclusively by humans and bacteria, suggesting they have been incorporated specifically into our genome.[5] Our guts are packed with bacteria whose aggregate microbiogenome has about 150 times more genes than "we" do, 3.3 million to our 23,000.[6] But *they,* though they come and go more easily than the rest of us, changing our mood and food, are also *us.* The immune system itself seems to be an evolved metasystem, a convoluted go-between, marshaling regulation and inflammation, and making sure that our animal cells and the rest of us—our bacteria and archaea—take it easy on each other.[7]

Hypersex is a provisional name for the commingling of organisms that meet, eat, engulf, invade, trade genes, acquire genomes, and sometimes permanently merge.[8] Life displays mad hospitality. The Korean biologist Kwang Jeon of the University of Tennessee received in the 1970s a batch of amoebas infected with a deadly bacterial strain. Most died. In a set of careful experiments after culturing the survivor amoebas for several generations, he found that the survivors, with fewer bacteria per cell, could no longer live without their infection. Deprived of their new friends and former enemies, the nuclei would not function without micro-injections of bacteria into the cytoplasm. The sickness had become the cure; the pathogens had become organelles; the last had become the first.

Had Jeon, who was a Christian, witnessed speciation in the laboratory? It seems so. But it was not gradual, as neo-Darwinism predicts. It was near-instantaneous, the result not of mutations accumulating in a lineage but of transformative parasitism.

SYMBIOGENETICS

Peculiar behavior, you say? Not really. Considering that life has been growing on Earth for some 3.8 billion years, it is not surprising that life has grown into itself, eaten itself, and merged with itself. Crowd control has long been an issue. Radical solutions have long been the norm. In 2006 researchers at Texas A&M University and the University of Glasgow Veterinary School in Scotland reported in the *Proceedings of the National Academy of Sciences* that endogenous retroviruses called enJSRVs are essential for attachment of the placenta and therefore pregnancy in sheep. We are as pure as the driven slush.

Like bacteria, viruses "R" us: They have moved into our genomes. Viral structural proteins have been "hijacked" and integrated into mammal reproductive tissues, immune systems, and brains. Some retroviruses disable receptors that lead to infection by other retroviruses. There is no racial, let alone genetic, purity in life. At bottom we are part virus, the offspring not just of our parents but of promiscuous pieces of DNA and RNA. The road to humanity is paved with genetic indiscretions and transgressions, no less than sheep would not be sheep without their acquired enJSRV.

The symbiosis expert Margaret McFall-Ngai asked a roomful of doctors what it meant for our marine ancestors to be surrounded by all those germs—about a hundred million cells per liter. They had no answer, but she told them: She has proposed that the immune system evolved not to eliminate pathogens but to select for symbionts in the microbe-packed waters of our metazoan ancestors.[9] The immune system in its origin may thus be more like an employment agency, recruiting desired species, than like a national security state, recognizing and refusing entry to guard the fake purity of the Self.

Today it is widely recognized that the cells of animals were once a wild party of two if not three ancient beings: the oxygen-poisoned archaeon host, the oxygen-using bacteria that became mitochondria, and perhaps wildly squirming spirochetes, which abound in anaerobic environments. These wrigglers often penetrate their fellows, which have no immune systems. They feed at the edges, becoming snaky motors propelling their brethren, or take up residence inside them, wiggling happily ever after.

According to my mother, who's been right before,[10] ancient bacterial symbioses gave our ancestors the intracellular motility abilities we see in mitosis and in the growth of undulating appendages. The creation of new symbioses by mergers on a crowded planet is called symbiogenesis. And we might call all aspects of its study "symbiogenetics"—the science of normative symbioses, the word commanding respect because of its apparent coinage from genetics; in fact, I derived it directly from symbiogenesis, though the connotation is a good one. Although this type of evolution sounds bizarre—a monstrous breach of Platonic etiquette in favor of polymorphous perversity—it is now confirmed by genetic evidence, taught in textbooks. It is a fact, or what the French philosopher of science Bruno Latour and the Belgian physicist-turned-philosopher Isabelle Stengers, not putting too fine a point on it, would call a factish. Nonetheless, although symbiogenesis—the evolution of new species by symbiosis—is now recognized, it is still treated as marginal, applicable to our remote ancestors but not relevant to present-day core evolutionary processes.

This is debatable. We are crisscrossed and cohabited by stranger beings, intimate visitors who affect our behavior, appreciate our warmth, and are in no rush to leave. Like all visible life-forms, we are composites. Near unconditional hospitality is necessary when we consider the sick factish that most of the human genome may be viral DNA.[11] Lactating women transfer their six hundred species of bacteria to their babies, as well as oligosaccharides their babies cannot digest but that feed certain bacteria. Bruce Sterling writes science fiction about humans engineered not to have any bacteria, but experiments with real mice deprived of their bacteria developed abnormal levels of immune system cells called invariant natural killer T cells that turned on their hosts, causing higher levels of inflammation, asthma, and inflammatory bowel disease. Although we are not mice, human studies show that early exposure to antibiotics is associated with asthma. The idea that we need to be pure and free of microbes to be healthy is as medically misguided as eugenicist dreams of triumph through racial homogeneity. Hundreds of species of fungi live in mammal guts despite or because of our immune systems. Indeed, scientists found that immune cells produce dectin-1, a protein that feeds skin fungi of mice; when they engineered the mice not to produce it, the mice experienced tissue damage from excess inflammation. It appears that our immune systems are designed not just to get rid of dangerous strangers but to entice needed others. No notion so disrupts the Pasteurian meme of health through biotic purity as the interest recently generated in fecal transplants, which have been declared safe in treating overgrowth of *Clostridium dificile* in patients needing to restore gut biota devastated from antibiotics, and are being investigated for the treatment of obesity. Another new medical approach is to develop skin creams that feed beneficial bacteria, warding off pathogens like *Staphylococcus aureus.*[12]

Of course these are medical avenues developing within a disease industrial complex that has long been in an emergency mode and makes its money from treatments. But symbiotic partnerships have the possibility not just to restore health but to improve it and alter us, to evolve us into new forms. I can envision future people with harmless luminous patches of bacteria, like tattoos but glowing in the dark, perhaps responding to mood or flashing like fireflies.

Some partnerships are fantastic. Luminous bacteria cram together to provide various marine animals with organs to light their way; deep-sea anglerfish females even use their shiny bacteria lights as lures to catch other fish. Luminescent bacteria, of the species *Vibrio fischeri,* provide the bobtailed squid, *Euprymna scolopes,* a nocturnal animal that feeds in the moonlight, so-called counterillumination: it projects light downward from its light organ, so it doesn't show up as a tasty morsel outlined in silhouette for hungry predatory fish below.

Nestled within the chromosomes of some parasitic wasps lie bacteria. Multiple insect species transform because of *Wolbachia* bacteria. The genus is nearly ubiquitous in insect tissues. Too big to fit within the sperm of insects, infective *Wolbachia* can confer parthenogenesis on insect populations, that is, transform a population with two genders into one that is all females, this of course to the advantage of the "selfish" bacteria, as the sperm bottleneck impedes their propagation. By disabling the gender-bending bacteria, antibiotics can make separate species of jewel wasps interbreed again. More bizarre than the space aliens we imagine abducting and toying with us on their saucers, these gender-changing bacteria bring in suites of genes for metabolic and reproductive features as they establish symbioses, often permanent, in arthropods.

WEIRD DALLIANCES AND UNEXPECTED SPECIATIONS

In an unexpected textbook example of speciation, the Columbia University geneticist Theodosius Dobzhansky selected fruit flies for their ability to withstand heat and cold. Dobzhansky found that after two years the heat-adapted flies could no longer successfully fertilize cold-living ones. The two separated populations of *Drosophila paulistorium* now conformed to the traditional zoological definition of new animal species. They had been reproductively and geographically isolated, and were now only able to breed with their own kind.

However, Wolfgang Miller of the University of Vienna Medical School, Austria, later found that the "cold-fertile fly population" had retained a symbiont widely distributed in certain tissues, whereas the "hot-fertile flies" had been "cured" of the symbiont. In fact, Dobzhansky's flies evolved as a result of the presence or absence of "mycoplasma," now recognized to be in the aforementioned genus of *Wolbachia.* In other words, the presence or absence of a bacterium, not neo-Darwinism's much-vaunted but still theoretical gradual accumulation of random genetic variations, correlated with what is, besides the Jeon experiments, perhaps the only real-time observed example of a speciation event.

Humanity's discovery of and battle with pathogenic microbial strains has misled us to think that microbes are generally extraneous to our bodies and health. But increasingly scientists are realizing that we have, we in part are, an adaptive microbiome, the endogenous collection of often "smart" microbes that is not only negatively connected to sickness but positively correlated to warding off obesity, asthma, allergies, and other maladies. Tel Aviv University's Eugene Rosenberg found that *Drosophila pseudoobscura* fruit flies would mate only with others on the same diet; antibiotics removed their dietary pickiness, leading to promiscuity and suggesting that possession of specific gut microbes can be like membership in a special club, leading to selective mating and ultimately speciation. And it is likely not just insects. Our symbiotic bacteria are connected to digestion, sense of smell, immunity, and other aspects of physiology. Prokaryotes are part of the hologenome; they are not just hangers-on but genetic actors.[13]

Such are the new facts—factishes—of life. As genes are not selves, the notion of the selfish gene remains a trope. Selves are materially recursive beings with sentience, and the minimum self seems to be a cell. Because life is an open thermodynamic system, as well as an open informational one, genomic transfer is rampant.

Leaflike green slugs (recently shown to manufacture chlorophyll themselves)[14] and underwater snails with rows of green plastids feeding them show how plants and animals can merge. *Convoluta roscoffensis* does not eat but burrows under the sand of the beaches of Brittany out of harm's way when the surf pounds (or a research scientist stomps his wading boots); when the danger passes, the animated algae, the green worms, then reemerge into the sunlight. The "planimal" is fed by the green building of its body, the living architecture that it gardens and which feeds it from within.

It seems unlikely that any cosmic deity arranged for the partners that are *C. roscoffensis* to come together, but they did, partly of their own accord, and they probably would have looked odd anyway on Noah's Ark.

Identifiable new behaviors, combined skills and physiologies, and even multigenome personalities also affect us. Human gut microbiota are not simply hangers-on but influence the timing of maturation of our intestinal cells, our internal nutrient supplies and distribution, our blood vessel growth, our immune systems, and the levels of cholesterol and other lipids in our blood.[15] They also—partly because of the presence of neurons in the mammalian intestinal tract, and the communication between gut and brain—influence human mood. Lab work with *Campylobacter jejuni* shows that this bacterium increases anxiety in mice, whereas the soil bacterium *Mycobacterium vaccae* inside them cheers them up. In people it has been suggested that yogurt with live cultures, for example with bifidobacteria, improves our sense of well-being.

TOXO

Toxoplasma gondii is a protist notorious for infecting pregnant mothers who may contract it from kitty litter. From the mother *Toxoplasma* moves to the fetus, often devouring it and leading to a miscarriage. *Toxoplasma gondii* sexually reproduces in bodies of members of the Felidae family, notably house cats. But mice, usually afraid of cats, lose their fear when their brains become infected with *Toxoplasma.* Also, and curiously, they become sexually attracted to feline urine.

Toxoplasma also infects large numbers of humans. Even though a possessor of *Toxo* may not know it's there, he or she can be affected by it. *Toxoplasma* infection in men correlates with enhanced risk taking and jealousy. An index of the risk-taking behavior is provided by the fact that males in car and motorcycle accidents are more likely to test *Toxoplasma*-positive.[16] *Toxo*-men are more likely to be unfriendly, unsociable, and withdrawn. Even if bereft of obvious symptoms, men who carry *Toxoplasma* are less likely, relative to controls, to be found attractive to women.

Women are another story. Women with *Toxoplasma* are more likely to be judged outgoing, friendly, and conscientious—and promiscuous. There is, of course, the complication of the stereotype of the woman who lives alone with all the cats. The caricature of such a woman is not of someone outgoing. Perhaps *Toxo*'s effects alter with age.

However confounding, *Toxo*'s effects seem real. *Toxoplasma* makes enzymes (tyrosine hydroxylase, phenylalanine hydroxylase) that alter brain levels of the neurotransmitter dopamine. Dopamine is a neurotransmitter involved in attention, sociability, and sleep. Cocaine and amphetamines work in large part by blocking the reuptake of dopamine in the brain. Dysfunctional dopamine regulation is theorized to be linked to schizophrenia, and several antipsychotic drugs target dopamine receptors. Up to one-third of the world population is thought to be infected with *Toxo,* with an estimated infection rate of almost 90 percent in France—a result perhaps of their love of rare beef, steak tartare or *saignant,* "bleeding." More alarmingly still, the Czech scientist Jaroslav Flegr found via MRI scans that twelve of forty-four schizophrenia patients showed significant shrinkage of the cerebral cortex, but that the reduction in gray matter of the schizophrenics was almost completely correlated with those who tested positive for *T. gondii.*[17] *Toxo,* accounting for a range of effects and affects from increased sexual attractiveness and feelings of well-being to full-on mental dysfunction, appears to be a facultative part of our more-than-human hologenome.

We have other "inner aliens." *Candida albicans* is the yeast fungus that causes vaginal infections and perlèche, a cracking at the corners of the lips. It thrives on easily digestible sugars and carbohydrates such as those found in beer, wine, cracker crumbs, and confections. It was perhaps spread among the wine-drinking revelers of Provence, troubadours who sang and jested, and who may have used makeup as a way to cover cracked lips that literally hurt when they smiled.[18]

Spirochetes are a stranger case still. Disease species cause Lyme disease and syphilis, and also other conditions. Spirochetes can go into hiding and form "round bodies," becoming virtually undetectable in cells. Friedrich Nietzsche and others are thought to have been infected by syphilis, whose "tertiary stage" is sometimes marked by a strange clarity of expression and artistic genius as well as madness.

ALFRED NORTH WHITEHEAD ON FACTS

I have talked about how the "facts" of symbiogenesis can in some sense be considered superior to the theory of neo-Darwinism. But since I am speaking about scientific facts to anthropologists,[19] I should probably be careful, as there is always the possibility that I am projecting cultural ideas onto the data and that all that we see or seem is but a culturally refracted dream.

According to Alfred North Whitehead, science is the bastard offspring of "irreducible and stubborn facts" (a phrase he took from William James in a letter to his brother, Henry) and the Greek genius for lucid theorizing.[20] Whitehead argues that, far from mental gymnastics, an "anti-intellectualist" strain was crucial for science's development—to protect it from the insular hyperconceptualization of mere academic thought.[21]

Science had to move away from the mannerist overdeveloped rational architecture of philosophy. While the Greeks had developed a remarkable ability to think boldly and clearly, and proceed through precise logic, the medieval scholastics, following Aristotle, expanded reason into a self-perpetuating empire out of touch with the real world. And the antidote was not more thinking but engagement with the real world. Of course, for natural science, engagement did not mean observation of other people, their thoughts or practices, but rather of things and organisms.

Whitehead traces this antischolastic attentiveness, which first developed among some Europeans but belongs to anyone who will have it, in part to, of all things, Greek tragedy—whose essence he says is "not unhappiness [but] the remorseless working of things. This inevitableness of destiny . . . This remorseless inevitableness," which in human dramas "involves unhappiness" but which "pervades scientific thought."[22] In short, for Whitehead the tragic realm of cause and effect has, if not a happy ending, a promising development: the development of modern science.

Closely observed by attention to facts, the inner workings of fate become reformulated as the laws of physics.

Interestingly, the Greeks—indeed the same Greek, Empedocles—came up with both symbiogenesis *and* natural selection thousands of years before Darwin. Empedocles had this great idea: In prehistory, organs, on their own, roamed the earth and recombined with one another. In other words, they symbiotically merged and were naturally selected. Those that persisted made copies of themselves, messily evolving.

Although Aristotle dismissed Empedocles because his mixed beings suggested irrational Greek myths, and Darwin dismissed Aristotle because Aristotle lumped natural selection with Empedocles, in retrospect

Empedoclean biology looks good. If you substitute cells for organs, Empedocles intuited both natural selection and symbiogenesis.

Alas, he did not engage the empirical. To Aristotle, the wacky misbegotten organs that arose on their own and coalesced to make bodies, the less fit ones dying out, must have smacked of passé myth, the mating of Olympians and humans, chimeras and immortals. For Aristotle, Empedocles's Dionysian imagination must have seemed a return to chaos, with no respect for observation or classification. The fifth century BCE philosopher of Acragas (now Sicily) was loopy. And for Darwin, who knew Aristotle was aware of natural selection because he mentioned it dismissively in connection with Empedocles, both Empedocles and Aristotle were wrong.

Yet as the spiral of science turns, what was once recognized as myth sometimes becomes re-cognized as science. This is the case with symbiogenesis. Strong evidence exists that all eukaryotes evolved suddenly by symbiosis and that many other organisms such as lichens, which combine fungi and algae or fungi and cyanobacteria, did also. We speak of brotherhood, but maybe we should speak of "otherhood." Others come together in aggregates of expanded energy use, economies of scale, and diverse assemblages where combined skills and redundancies prefigure additional developments. Together mingling beings find energy and the substances they need to live. Corals reefs require photosynthetic symbionts. "White ants," or termites, cannot digest wood without the living hordes in their guts whose visualization Joseph Leidy compared to the outpouring of a "crowded meeting-house." Cows, "four-legged methane tanks," collectively add enough of that gas, unstable in the presence of oxygen, into the atmosphere that aliens, outfitted with spectroscopic devices, might be able to tell that there was life on this planet from the presence of microbially produced methane alone. And that's no bull, though it's close, literally. (See chapter 10.)

It is hilarious to contemplate that the methanogens releasing gas in the specialized symbiotic cow stomachs called rumens could signal—with no help from humans—the presence of complex life on this planet. But this point that the archaean nonequilibrium chemistry of our atmosphere could serve as a beacon to aliens is serious, too. And it serves as a nice segue into related facts of life I want to talk about now.

SENTIENT FLAMES

Let me turn now briefly to what I consider another evidence-based discourse, a new set of life—and death—facts that may be even less well-known to you than symbiosis but which I consider equally important to understanding life.[23]

I don't think it's possible to understand life without understanding the role of energy. Life is a complex thermodynamic system. Like a whirlpool or flame, the shape alone stays stable as energy is used and matter is cycled. It absolutely depends on the energy and matter. Deprive a storm system of its atmospheric pressure gradient, an autocatalytic chemical reaction of its chemical gradient, or convection of its temperature gradient, and these disappear: their form is dependent on their function—not necessarily their only function, but their basic one. And the same is, elegantly, true of life: if you deprive cyanobacteria or purple sulfur bacteria or plants of their solar energy gradient, or animals of their food-oxygen gradient, they disappear. The big difference with life is that it has found a way, via the recursive DNA-RNA-protein system, to restart the flame.

It is a thermodynamic fallacy that we are destined to die because of wear and tear and inevitable entropic dissolution connected to thermodynamics' second law. In fact, life's signal operation, as well you know, is to resist normal wear and tear. While writing this essay, I saw a little kid's face light up when his parents pointed at a Starbucks and pronounced the magic word, "cookie." He recognized that sugar gradient much as a bacterium swimming toward sweetness or a sunflower following the light. Of course he is much more evolved than a bacterium or a sunflower: He is human. More to the point, taking a break from writing this essay, in the shower, I rubbed my eye and it swelled and reddened something awful. "Miraculously," however, it restored itself.

Finding food to support its body and using energy to repair itself, which occurs even at the DNA level, are typical operations of life. But what is life doing in its cosmic context? I would argue, and have, that the metabolic essence of life is to degrade gradients. Inanimate complex systems do this, but staying alive prolongs the process. I consider this a lucid, Greek-style idea. And there are facts to back it up.

Apoptosis, telomerase-based limits on cell divisions, and sugar- and insulin-mediated genetic mechanisms ensure aging-unto-death in most familiar species. But locusts, mayflies, and other organisms that "come unglued" (experience multiple failure of organs), dying within hours of reproduction, contrast greatly with long-lived ones like sharks, lobsters, and some turtles not known to age. (Sharks, who often devour their twins in utero, may not need to die, as they are exposed to so many death threats.)

That aging is under genetic control is attested to by the difference between Pacific and Atlantic salmon, the latter of which return upstream for another bout of egg laying. The energy connection here is that populations that grow too rapidly without moderating their growth run the risk of being wiped out by famines and epidemics.[24] Death by aging, in other words, is not an accident but an adaptation. The important datum that near-starvation is the surest life-span extender in organisms as distant as apes and yeast (evolutionarily separated by some seven hundred million years) suggests that hunger acts as a signal to slow down aging programs, thereby increasing the chances of population and species-level survival. The modulation of aging in the face of environmental signals of scarcity is an example of physiological prudence among cells and compares favorably with conscious human attempts to moderate population growth.

IN ADDITION TO THE SEEMING GENETIC FETTERS on unrestricted growth in our own bodies, and in populations of aging animals, consider the actual work done by plants. We like to think our symbol making and technology make us superior, but plants are metabolically superior in that they can derive energy from oxygen, which they do at night when sunlight is not available as a source of energy. They can switch-hit like this because they also incorporate those former respiring bacteria, the mitochondria, into their cells along with the plastids, which we never got, making us worry about where to get the next meal.

And plants are far from inert. The average condensation and precipitation from soil and leaves in midlatitudes during the summer is about six millimeters per hectare. This production of latent heat via evaporation off of leaves is the energetic equivalent of some fifteen tons of dynamite per hectare. Rainstorms are far more energetic. Seeded by evapotranspiration from trees, rainstorms release the equivalent of many megatons of dynamite. But they do it stably. Competing with one another for access to the light that drives their evapotranspiration, trees disperse more energy more steadily than we do even with all our technology.[25]

We think we are smarter, but in the long term we haven't proved ourselves to be. Indeed, we may be heating up the planet, which is a clear symptom of dysfunction in complex systems. Think of your laptop, overheating. Natural complex systems use energy and dissipate it elegantly and have learned to do so in stable ways over millions of years of evolutionary time. It is true that we are Promethean, gifted in our ability to locate and exploit energy gradients. It even happens in our own bodies, whose brains use 40 percent of our blood sugar to spin forth fancies of variable value. But the long-term thinking we pride ourselves on is not in evidence when you consider thermal satellite evidence that rainforests are the most efficient coolers of the planet. These biodiverse collectives naturally use energy, but they dissipate it away from their surface, and do so sustainably.

In *Alien Ocean: Anthropological Voyages in Microbial Seas,* Stefan Helmreich writes: "At the conclusion of *The Order of Things,* Foucault, in a phrasing that evokes Rachel Carson's description of the seashore world, suggested that *man* may someday 'be erased, like a face drawn in sand at the edge of the sea.' He did not mean," he adds, "that humanity might be wiped out by oceanic inundation—though such a literal reading is freshly thinkable . . . in the wake of the Indian Ocean tsunami of 2004, 2005's Hurricane Katrina, and growing evidence of global warming. Rather, Foucault speculated that the *human*—that biological, language-bearing, laboring figure theorized by human sciences ranging from anatomy to anthropology to political economy—might not endure forever, just as archangels, warlocks, and savages are no longer so thick on the ground of our social imagination as once they were, and just as *race* as a biological category now wobbles between phantom and Frankenstein as it has been set afloat in a sea of genes."[26]

I believe anthropology's new engagement with the nonhuman may be another example of "the return of the scientific repressed," but I believe it also represents increasing pressure on us to become more integrated into more biodiverse, energetically stable ecosystems. Populations tend to be most numerous in the generations prior to their collapse.

Stem cells and pioneer species spread rapidly but become integrated in slower-growing adult organisms and ecosystems that optimize and sustain energy use. In this light, humanity as a whole seems to be ending the insular rapid-growth phase typical of immature thermodynamic living systems. This view provides a possible new positive interpretation of Franz Kafka's witty lament, "There is hope, but not for us."

CHAPTER 2

BATAILLE'S SUN AND THE ETHICAL ABYSS

Late-Night Thoughts on the Problem of an Affirmative Biopolitics

> Nazism treated the German people as an organic body that needed a radical cure, which consisted in the violent removal of a part that was already considered spiritually dead. From this perspective and in contrast to communism (which is still joined in posthumous homage to the category of totalitarianism), Nazism is no longer inscribable in the self-preserving dynamic of both the early and later modernities; and certainly not because it is extraneous to immunitary logic. On the contrary, Nazism works within that logic in such a paroxysmal manner as to turn the protective apparatus against its own body, which is precisely what happens in autoimmune diseases.
> —*Roberto Esposito,* Bíos

TODAY IS THE FIRST DAY of the rest of your strife. In thinking about ethics we come up against some of the most difficult problems. One person's righteous indignation is another's reactionary oppression. The citizen's free speech can be the government's hate speech. The model's sexy furs are PETA's incontrovertible evidence of animal slaughter. Your nice iPhone may entail child labor, environmental degradation, and a Chinese worker's exploitation. Even the seemingly innocent sweep of a linoleum countertop may represent, from another level, microbial genocide. When this example was brought up before a roomful of students in Danville, Kentucky, in the context of a discussion of life's extent in the context of, among other things, abortion, many in the class raised their hands when asked if they believed microorganisms were not alive. For a sperm and an egg cell, fertilized or not, do not look that different from many microbes. Do they have feelings? Is the male masturbator guilty of wanton destruction of human life? The vegetarian (and Adolf Hitler was one) may think eating meat is murder, but thinks nothing of flying to an environmental conference, thereby adding to global warming that may trigger a wholesale climate collapse. Still others would argue that wiping humankind off the face of the Earth in the long run may be just what the biosphere needs to keep going.

Somewhere Jacques Derrida writes that all of his work amounts to nothing but graffiti on the base of the monument that is the work of the rabbinical religious thinker Emmanuel Levinas. A friend of Maurice Blanchot and who at first admired Martin Heidegger, Levinas last century recognized that there is no possibility for a prescriptive ethics, a Mosaic tablet of writs set in stone that will guide us as to conduct, what is right to do, as we make our way through the ethical darkness. We need instead a descriptive ethics; we must engage with the other as other, falling without parachute through an abyss without bottom. The pointing finger has three fingers pointing back at the accuser. We have moral dyslexia. Who can guide us? For Levinas, it may be God, or what is left of him after Friedrich Nietzsche. We must be there with the face of the other, accountable, responsible to it. I read large sections of Levinas's *Totality and Infinity,* and I thought this idea of grounding ethics in the face, before the face of the other, was a fascinating idea. I was in a rush to apply, to appropriate it. I did so, naively no doubt,[1] to the face of Earth so that we might have ethical accountability toward the planet, our mother, the biogeochemical matrix from which the flesh of our body comes but also the environment we co-opt and infect in our nonstop proliferation. There is a cool painting that shows a lunar-landed astronaut, the blue Earth reflecting off his visor, obscuring and replacing his face. The Levinasian ethics of the face also seemed to touch on the lack of accountability in technowarfare, dropping bombs on those we don't see, death at a distance. Recently, however, I learned (during a lecture by Cary Wolfe, the editor of the University of Minnesota Press's Posthumanities series) that Levinas didn't even consider animals to have a face. That was strange. Animals don't have a face? Dogs have no faces? What kind of a face is this? And I'm no Bible scholar, but I can't help think what I've heard—that, although God never shows his face in the Bible, there is a key passage in which he flashes his backside.

IF NATURE IS AMORAL and religion offers us no reliable moral code, where do we go for our ethics? In reaching for an affirmative biopolitics, I want to talk about the productivity of the agon and make a few probably unpopular comments about what we call war and the general climate from which it comes. I believe we are in a dysfunctional relationship with Big Brother, and it is nonconsensual. But I don't agree that the strife that prevents an affirmative biopolitics can be laid simply at the foot of mononaturalism, as Bruno Latour argues, or that its roots are simply human.[2]

On the other hand, I think that facing the deep roots of what emerges as violence in the human realm can help us understand if not address it, and that this has a Spinozistic–Madame Curie virtue beyond activism. Curie said that nothing in life is to be feared, only understood; Spinoza argued that, with free will an illusion, true freedom lies on the way of knowledge.

It is true that contingent human history shapes what we take to be universal scientific knowledge, but this contingent human history also reflects the larger thermo-cosmic evolutionary situation in which we are embedded. I live in a world in my head in the world, said Paul Valéry. We dwell in a nature circumscribed by culture inside nature. Whether that second nature is also inside culture I'll leave for you to decide.

I WANT TO MENTION quickly what I think is the basis of the ethical problem of life on Earth. It is twofold. First, sensing, sensation, including the avoidance cues of pain, which we may assume is among the oldest phenomenologically detectable signals, correlates with living beings. Second, Earth is essentially a materially closed thermodynamic system. Like other natural complex thermodynamic systems, material cycles as energy flows, and the system, if possible, grows to use up available resources. Microbes have mastered complete recycling of chemical elements in ecosystems. But if we look at evolutionary history, there comes a time when organisms developed the potential to consider themselves individual selves. I would provisionally locate this potential chronologically with the Ediacaran fauna, among the first organisms to have heads. These beings lived earlier than the trilobites. They may not have been animals at all, but symbiotic organisms living with algae in their tissues. But either they or the animals that followed them recognized each other as gradients. This set up an ethical crisis. Animals not only devour each other—"meat" is the name of that gradient—but their perception and intelligence allowed them to hunt. This is the same awareness that would ultimately allow us to know we harm others to feed, and that someday we will die.

Thus somewhere in the evolutionary history of animals, after they diverged from fungi some seven hundred million years ago—and according to James Watson, 40 percent of yeast proteins are still homologous to ours—there came a point where, with their sensory organs concentrated at one end, they recognized their fellows as a rich energy gradient. Of course they didn't at first realize, as we vegetarian, vegan, Jainist, and pepperoni pizza eaters do, that those fellows were feeling beings like themselves. But this primordial carnophagy, as Derrida calls it, set up the conditions for an ethical crisis from which we still have not recovered.

We come from a long line of naturally self-centered ancestors. As Alan Watts, the greatest popularizer of Eastern philosophy we have ever had in the West, puts it, the "shape alone is stable. The substance is a stream of energy going in at one end and out at the other." The tubes "put things in at one end and let them out at the other. . . . [This] both keeps them doing it and in the long run wears them out. [But this part isn't true, as I will explain.] So to keep the farce going, the tubes find ways of making new tubes, which also put things in at one end and let them out at the other. At the input end they even develop ganglia of nerves called brains, with eyes and ears, so that they can more easily scrounge around for things to swallow. As and when they get enough to eat, they use up their surplus energy by wiggling in complicated patterns, making all sorts of noises by blowing air in and out of the input hole, and gathering together in groups to fight with other groups. In time, the tubes grow such an abundance of attached appliances that they are hardly recognizable as mere tubes, and they manage to do this in a staggering variety of forms. There is a vague rule not to eat tubes of your own form, but in general there is serious competition as to who is going to be the top type of tube."

All this seems "marvelously futile," says Watts, adding that it is "more marvelous than futile."[3] Watts is right, although it is actually a bit worse from an ethical standpoint because his assumption that the tubes must wear out turns out to be wrong. In fact, they are killed off by what we could call an "inside job": multiple redundant systems, working from deep within the genome, ensure that organisms die by aging in many species, including us. Genetic assassins—to use colorful language—include apoptosis (programmed cell death), telomere shortening, and glucose-mediated mechanisms that ensure we age relatively quickly if we are well fed.[4]

Of course to say "assassin" and "inside job" is to beg the biopolitical question. For example, Ed Cohen in *A Body Worth Defending* excavates the political roots of the seemingly self-evident idea of the biological idea of immunity. And he does a splendid job, beginning his book with the striking image of Élie Metchnikoff, who introduced the term *immunity* into biology. One day Metchnikoff, when his "family had gone to a circus to see some extraordinary performing apes, remained alone with [his] microscope, observing the life in the mobile cells of a transparent star-fish larva [at which point] a new thought suddenly flashed across [his] brain."

As the scientist described his finding, "It struck me that similar cells might serve *in the defense of the organism against* intruders . . . I said to myself that if my supposition was true, a splinter introduced into the body of a star fish larva, devoid of blood vessels or a nervous system, should soon be surrounded by mobile cells as is to be observed in a man who runs a splinter into his finger. This was no sooner said than done.

"There was a small garden to our dwelling . . . [and] I fetched from it a few rose thorns and introduced them at once under the skin of the beautiful star-fish larvae as transparent as water.

"I was too excited to sleep that night in the expectation of the results of my experiment, and very early the next morning I ascertained that it had fully succeeded."[5]

COHEN CORRECTLY POINTS OUT that Derrida before he died tended to speak of 9/11 in terms of global autoimmunity. Cohen criticizes Derrida for conflating immunity and autoimmunity, but as I read, it occurred to me that this conflation may reflect our own geopolitical confusion. Whether or not Derrida was aware of insufficiently investigated physical anomalies on September 11, 2001, his conflating immunity and autoimmunity acts like an unwritten Wiki entry that can absorb future discourse. Speaking plainly, if 9/11 is indeed a false flag (one of many and not necessarily the most recent in a long military history), then Derrida's "conflation" is quite cagey.[6] Immunity would refer to the rapid spread of national security control apparatuses in the global body politic, whereas autoimmunity would refer to the purposeful introduction of terror, a state crime against democracy, to initiate a global response.[7]

As I read I also worried about what we might call Foucauldianism or even Foucaultitus (pronounce it how you like)—the tendency to pore through documents and identify a concept or reality with its historical introduction into texts. Sex began in such and such a century, immunity evolved in the court system, and so on.

As Cohen relentlessly pushed for the political, social, and cultural roots of this seemingly innocent concept of immunity for almost three-hundred-plus pages, I kept asking myself, yes, but if the starfish, and we, are not in some sense immune, what would you call it? What is the transparent starfish larva doing with its mobile cells extracting the rose thorn if not being in some sense immune to them?

THERE IS OBVIOUSLY SOMETHING to be said for this modern form of scholasticism, as it forces us to disinter concepts we take for granted. However, I believe Alfred North Whitehead had it right when he credited a deep strain not of conceptual gymnastics but of "anti-intellectualism" to the brilliance of science.[8] The excesses of Aristotelian scholasticism served as a counterexample, helping scientists leave the musty room of involuting ideas with no love lost, as they stepped forward with just the Greek genius for bold lucid speculation, into the fresh air of nonhuman things, which they measured and observed. . . . A good example, of course, is Metchnikoff himself boldly applying the juridical concept of immunity to starfish larvae, and finding that it worked.

By book's end, however, Cohen answered, or acknowledged, my question. He discussed AIDs as an insufficiently problematized diagnosis, one dependent on the very notion of immunity, which provides the

"I" in the middle of both AIDS and HIV. "Might biological *community*," he writes, "enable us to appreciate healing. . . . There may be more to immunity than we currently know, or are indeed even capable of knowing, so long as we remain infected by the biopolitical perspectives that it defensively defines as the apotheosis of the modern body."[9]

Here it is worth a return to Margaret McFall-Ngai's startling but sensible surmise that the immune system, appearing first in marine metazoans surrounded by seawater infused with one hundred million microbes per liter, started by not weeding out or destroying pathogens but engaging the most helpful symbionts[10]—producing, as Cohen suggests, a community.

STILL AND ALL THE SAME, metacaspases, T cells, nitrogen oxide, telomere rationing, apoptosis, and thymic involution (which progressively weakens the immune system) seem part of cell regimes that not only shape and protect but eventually kill the body via aging. While Latour recommends a constructivism in working for world peace because naturalism has been tried already, and didn't work, this recommendation runs the risk of derailing the sort of critical thinking necessary to see the depths of problems before beginning to fabricate novel solutions. We don't have to be mononaturalists to see that culture is grounded in a nature or that nature is not our organic fairy godmother.

Facing the problem of culture's natural roots squarely, Roberto Esposito writes, "Anything but the negation of nature, the political is nothing else but the continuation of nature at another level and therefore destined to incorporate and reproduce nature's original characteristics."[11] It is tempting but insufficient to argue that something like the general violence that war for us epitomizes is a contingent peculiarity of human history that we've prematurely naturalized in concepts like immunity and survival of the fittest. We need not paint with as broad a brush as Nietzsche: "Every moment devours the preceding one, every birth is the death of innumerable beings; begetting, living, murdering, all is one." But Spencer had a point. "Now that moral injunctions are losing the authority given by their supposed sacred origin, the secularization of morals is becoming imperative."[12] This paves the way for religionists to contemplate the moral vacuum that threatens them with "fear" and "dread." On the other hand, it emboldens political movers and shakers, not to mention sociopaths, to seize the opportunity in a world without moral law. However, I would add that it's much harder to break natural laws than human ones.

In the gruesome feast at the end of Zarathustra, Overman sets fire to his texts and dances. This is *aktiv Vergesslichkeit*—active forgetting. I think we actively forget when we are confronted with life's ethical abyss. The presence of this abyss reminds me of the Sun for Georges Bataille. In 1929 in Paris, in translation, Bataille read Vladimir Vernadsky's *La Biosphère.* Vernadsky, as famous in Russia as Darwin is in the West, argued that we are children of the Sun, that life is a solar phenomenon, that the Earth system is really the Earth–solar system. Accumulating light energy in green material through photosynthesis, life creates problems, especially for itself. I would liken life's ethical abyss to the Sun as described by Bataille: Although it is real, our great source of excess and possibility, we tend to abstract it or look away, because we cannot gaze on it too long without going blind.

CHAPTER 3

THE POST-MAN ALREADY ALWAYS RINGS TWICE

THE FUTURE TIMES

"Today" I received this strange news item, but it had no address on it, so I am passing it on to lucky you. By post-man I will have meant (mostly) posthuman. And he (so to speak) rings twice. At least.

The first ringing is literal and refers to what comes after humans in evolution. The first ringing announcement that the posthuman has arrived has to do with speciation, guesswork, machines; with loose predictions that fall off a cliff of accuracy as we extrapolate physically nonextrapolatable trends into the future. The classic example of such a trend is the graph of human transportation modes over time. I discovered this in John Platt's futurism class at Harvard, which I audited in the 1980s (after 1984, before 2001). Extrapolated, the hyperbolic curve of increasing velocity made by peripatetic pedestrian philosopher, horse-drawn carriage, automobile, airplane, and rocket ship does not take long to exceed the speed of light, a progression that is impossible under Albert Einstein's general theory of relativity. Another example—more pertinent and potentially horrible—is an extrapolation of human population growth. I committed the "fact" of total world population to memory in grade school. It was three billion. Now it's more than doubled. We know from the fossil record and observation of bacterial growth in petri dishes that the greatest species abundance sometimes occurs in generations immediately prior to population collapse. The first ringing of the posthuman is thus linear or literally futuristic, and as such speculative, most readily explorable by the science fiction imagination.

Unfortunately for us as humans, the majority of fictive posthuman realities are characterized by the total absence of Man, except of course as the imaginer. (I mean "Man" in the old, general, sexist sense of the term. Of course, some futures include men without women and women without men. In some futures there will be only women, cloned lesbians more satisfied and less ecologically destructive without the Y chromosome. Perhaps they will replicate Arnold Schwarzenegger's disembodied pectoral muscles for white breast-meat in annual Thanksgiving to their triumph. Nonetheless, my meaning of posthuman is inclusive.)

THE FUTURE UNTIMES

The second ringing or meaning of the posthuman is for me both more interesting and more accurate with regard to the posthuman, the after-Man. Because I take it as axiomatic that linear time may be (sometimes is) an illusion, the second, nonlinear or metaphysical aspect of the posthuman operates not in the science fiction future but in the present. Logically, of course, the future is a construct. As Saint Augustine of Hippo reportedly said, "I know what time is until you ask me." The future is always to come; we never get there except in our imaginations, as with the past. It is always now. Mystics and ecodelic-ingestors sometimes report the experience of seeing time "end-on," although the word *experience* here is inadequate insofar as it connotes precisely that duration in linear time that is being called into question. Alan Watts has stressed that while ecologists, biologists, and physicists *know* that the organism and the environment are not two things separated by a subject–object relationship—but are rather a single process, a unified field—they don't necessarily *feel* that this is so. This argument is similar to the idea that time is possibly illusory or insubstantial, a conclusion that can be logically arrived at by, say, asking you to point in the direction of the future. According to Benjamin Whorf, the Hopi locate time behind them, because it cannot be seen. The past is solid, visible, because we can see what happened (think of "hindsight"); the future is liquid, open, black, and frightening, but wonderfully full of potential. Outside time, such differences blend or further differentiate. Glossing Martin Heidegger in a footnote, Jacques Derrida describes how, in the nonvulgar or "Greek" conception of time, times past, present, and future converge and diverge; they are at once touching and infinitely distant. Perhaps the founding example of such posthuman nonlinearity is Friedrich Nietzsche's discussion of the last man and the *Übermensch* (the "Overman") in *Thus Spake Zarathustra.* The last man, the overcoming of man, and the superman are, clearly, not just literal (as in Adolf Hitler's interpretation) but allegorical. On the one hand, Nietzsche, thinking of Charles Darwin, complains how these English are no philosophers. On the other hand, Zarathustra—his prophet up in the mountains who talks to the animals and informs humanity that God is dead (but the news has taken a long time to reach us)—espouses the theory of eternal recurrence as his most important doctrine. At the very least, the idea that everything that can happen will happen, and not once but an infinite number of times, does not go along with simple, linear, evolutionary scenarios.

So we have then a double meaning, a double entendre, of the future: one linear, one nonlinear. The post-man's first ring announces the future of man; the second is uncannily silent, more a buzz or a beep. This is the human future that never arrives because it is already always here.

There are problems with both ringings. But in the end I think the nonlinear, atemporal or polymorphically temporal posthuman version is the more resonant. Ironically, the playful and metaphorical view of the human future is not only broader but more literally true than what Heidegger calls the received or "vulgar" view of orderly, progressive, linear time.

But before I veer off into unreconstructed mysticism, let me say that the fragmentary, the fallen, and the derivative do not belong only to the present age—as Derrida argues, against Jean-Jacques Rousseau, in *Of Grammatology.* Language itself, the splintering of things into signs, into alphanumeric representations, calls into question our idealized image of the past: the golden age, our happy childhood, or, cosmologically, the unity of the singularity at the big bang. This seems to be the same absent sense of wholeness, of the halcyon, that we project into the future: heaven, merging with light, a dissolution into the ecstasy of the sexual or the neural that relieves our constitutive sense of loss. Sigmund Freud's student Otto Rank located this loss in the original trauma of birth,[1] the loss of the womb—the experience of which might, if we could remember it, be compared with floating in space or lying on a raft on opiates in a sunlit pool. To be alive is to be deprived of this Edenic yesteryear for which we long, and this utopian future for which we strive. Jacques Lacan suggests that both are based on the illusion of the uncoordinated, wobbly toddler looking in the mirror, or at its mother, and hallucinating its own unity. Locating the mirror stage in the register of the Imaginary, Lacan marks it as the gateway into language, into the splintering of signification. The idea of wholeness in the past as well as in the future is thus a narcissistic illusion.

One familiar way to illustrate this doubleness, this insistent and split entity arriving at our threshold with the news that never quite comes but is already always here, is to look at radical science fiction scenarios for the far future and to see how they stack up against the present. A visitor from humanity's future is projected back in time, landing, for example, in lower Manhattan, in the middle of rush hour. "Ah, what peace!" he sighs. The witty geneticist J. B. S. Haldane said the universe is not only queerer than we suppose; it is queerer than we *can* suppose. Truth, or what Lacan calls the Real, is stranger than fiction, including science fiction. (Granted, not all agree. The fiction guru and Cornell professor Paul West scoffs at the idea that anything is stranger than fiction.[2] But I guess that, for me, fiction is a subclass of nonfiction, just as the linear is a subset of the nonlinear.)

There is a branch of science fiction concerned with the idea that life will evolve into robots, that consciousness will be based on silicon or metal rather than carbon. If an MIT graduate student could build such a silica-based robot—one that could make more of itself, a real nanotechnological von Neumann machine—he or she would get a Nobel Prize. But these nanobots already exist. We call them diatoms and radiolarians, and they build their beautiful tiny skeletons from silica, which they take from the ocean. With regard to metal, magnetobacteria have tiny magnets in them that they use to orient themselves to Earth's magnetic poles, and chitons scrape corals with their iron teeth. Life is ahead of the human game of fiction.

Or take Isaac Asimov's novel *The Gods Themselves.* Asimov depicts an alien world with three sexes, which join together at maturity into permanent union. Members of each sex are attracted to members of the other two. At the end they transform from ethereal beings that can move through each other into "hard ones" that are physically solid. In fact, something very similar has already happened in evolutionary history, when the ancestors of mitochondria penetrated archaea to make amoeba-like cells ancestral to your own. Another lineage was joined by cyanobacteria to make the cells ancestral to all algae, seaweed, and plants. Even the fluidity of bacterial movement, their ease of penetration and genetic transfer, was stalled with the evolution of the real "hard ones"—the eukaryotic ancestors to animals.

ANOTHER EXAMPLE of scientific truth being stranger than science fiction: In a story by the golden-age (1938–1946) science fiction author A. E. van Vogt, two spacecraft rendezvous. They link hulls, breed their human inhabitants, and, having accomplished their mission, continue on their interstellar way. Literally, this is a far-futuristic posthuman scenario. The spacecraft has become the skin or exoskeleton of the human. The mode of reproduction has radically altered, and the phenomenology of human existence, too. How strange. Yet this is, I would argue, an allegory for what has already happened to people. When we look at these lonely, far-future descendants who have forgotten their past but occasionally link up for some quality time in the depths of space, we have to think of ourselves. We are constructed partly of hard parts (e.g., the skull encasing cranial software) and have "forgotten"—lost touch with—the deepest parts of our biological history. The events that induce us to seek each other out and regenerate new beings that grow from amoeba-like cells are part of a bizarre microbial past compared with which our ape heritage seems positively flattering. Insofar as "we" remain cells, the "spaceships" of our bodies are at once glorious crafts carrying us across dry land and cyborg extensions of our soft and slippery inner core.

Moreover, we must assume that the Vogt crafts, in order to avoid needing to restock themselves with food, to be truly space-faring rather than merely "camping out" in space, they must be jointly a recycling, self-sustaining ecosystem. And if they continuously recycle bio-elements to support their symbiont-like humans, then we have not just an allegory but a literal description of present ecological relations on what

Buckminster Fuller called Spaceship Earth. A human being feeds collectively on its external ecosystem; one is not a Platonic island separate unto oneself. Thus the posthuman lesson from van Vogt's story (as I choose to interpret it, imagining the necessary details of the life-supporting craft) is that what we call "human" really includes the bacteria, the protists, the plants, and fungi—everything we need both within and outside our bodies to be what we are. The meeting of the human-containing spacecraft in the depths of space reinaugurates the pattern of interconnecting, thermodynamic flows, exposed only partly to conscious control, that established our cellular forms in the first place. We naturally gravitate toward repetition of those processes that allow for the genetically safeguarded repetition of our pattern of metabolic cycles in new, inevitably aging entities. It is instructive to think of machine and organic evolution together. Although, for example, we like to distinguish conscious mechanical creation from unconscious genetic evolution, modern devices such as color televisions depend for their manufacture on so many globally distributed processes that eventually these decentralized functions become self-perpetuating or "unconscious" at the level of individual worker or company. A woman does not make a baby consciously, yet her conscious choices can influence embryogenesis. Similarly, mechanical production processes that were once consciously engineered become more automated and "physiological" as consciousness, moving on to new and more pressing tasks, forgets its former triumphs.

The machine craft imagined by van Vogt are no less strange than our dependence on agricultural machines and cell phones, laboratory-made antibiotics, and factory-made automobiles. A true space-faring ecosystem requires not just ecologically complementary life-forms but also the mechanical "shell" necessary to protect life from astronomic insults, from radiation and collisions. We are open thermodynamic systems simultaneously using high-quality-energy sunlight (through the intermediary of food) to organize our environments, even as we inevitably degrade that energy, adding heat and entropy to our surroundings in a high-stakes process. Technology has appeared on Earth, while many species of mammals, and other animals and plants, have disappeared. We are already cyborgs packed with prostheses—although the prostheses are usually outside, not inside, our bodies. Yet even this may not be true. For, from a deep-ecological standpoint, we ourselves can be seen as spaceships made by massive populations of intermingling cells. Instead of an exoskeleton protecting us from cosmic rays, we have an endoskeleton protecting us from gravitational strain. Instead of sparsely colonizing space from a base on Earth, we celled organisms have densely settled the land from a base in the ocean. The calcium extruded by cells to make bones in our bodies was, for our ancient cell ancestors, a poison. Eukaryotic cells shed calcium ions, and this ancient marine waste was stockpiled; eventually it became involved in the metabolic processes of cell communication and growth, and evolved, in our animal ancestors, into muscle movement and, in us, into thought. The inevitable thermodynamic pollution resulting from life's growth was shaped by and incorporated into life's structure. And this appropriation of calcium was not an isolated incident. The same thing happened with free oxygen, which also began as a toxic waste. Over time, the outside becomes the inside, the house a body.

As we try to imagine the immense journey from cosmic birth through life's origin some ten billion years later to the evolution from bacteria of humans taking up the most recent third of known time, our temporal sense is strained. Nevertheless, our ability to imagine infinity, so important to mathematics and physics with its eternal laws, suggests a bizarre alliance between infinity and time. In evolution the post-man rings once, but in infinity he has already always rung. "The Earth is no more a rock with some life on it," it has been said, "than you are a skeleton infested by cells." The sheer temporal immensity of the evolutionary process may make it difficult, if not impossible, for us to picture it without Zen-like perspectival shifts that are apt to emphasize more what we don't or cannot know than what we can. Our hard parts and cells feel like one skin-encapsulated being—but when I think how, when I'm driving a car, a close scrape impinges on me physiologically, giving me a sensation as if the metal to which no neurons are attached were part of my own body, I am persuaded that, when it comes to ego, there is no absolute there there.

In other words, our bodies are what Samuel Butler, Darwin's contemporary and author of *The Way of All Flesh,* called microbial "toolkits." Butler envisions a day "when all men in all places without any loss of time are cognisant through their senses of all that they desire to be cognisant of in all other places, at a low rate of charge so that the back-country squatter may hear his wool sold in London and deal with the buyer himself—may sit in his own chair in a back-country hut and hear the performance of *'Israel in Egypt'* at Exeter Hall—may taste an ice on the Rakaia [a New Zealand river] which he is paying for and receiving in the Italian opera house. . . . [This is] the grand annihilation of time and place which we are all striving for and which in one small part we have been permitted to see actually realised."[3]

This is fascinating, for here, circa 1865, Butler—who took Darwin to task for making life too mechanical and got back at him by applying evolutionary theory to machines in order to consider them as natural—extrapolates the telegraph to anticipate the Internet.

Perhaps the greatest science fiction story would be a literal description of our present reality, but couched in terms that made it unrecognizable until near the story's end. We love the future because we don't know what it will be but because we can, to a slight extent, shape it and because in our narratives it is always uncannily familiar. In a linear frame, however, we have to be careful with prediction because it is too often simple extrapolation. Butler may have "predicted" the World Wide Web, but in decades past it was also predicted that every home would one day have its own rooftop helipad and that each town would have its own telephone. Linear extrapolation is doomed because exponential rates of change cannot continue. "Only two things are infinite," said Einstein, "the universe and human stupidity."

But rather than end with a mysticism-friendly scientist, let's give the "last" word to a science-friendly mystic. In 1974 Alan Watts wrote in an essay titled "Psychedelics and Religious Experience":

> The Western man who claims consciousness of oneness with God or the universe thus clashes with his society's concept of religion. In most Asian cultures, however, such a man will be congratulated as having penetrated the true secret of life. He has arrived, by chance or by some such discipline as Yoga or Zen meditation, at a state of consciousness in which he experiences directly and vividly what our own scientists know to be true in theory. . . . There is no way of separating what any given organism is doing from what its environment is doing, for which reason ecologists speak not of organisms in environments but of organism-environments. . . . The Western scientist may rationally perceive the idea of organism-environment, but he does not

ordinarily feel this to be true. By cultural and social conditioning, he has been hypnotized into experiencing himself as an ego—as an isolated center of consciousness and will inside a bag of skin, confronting an external and alien world. We say, "I came into this world." But we did nothing of the kind. We came out of it in just the same way that fruit comes out of trees.[4]

PART II
STARDUST MEMORIES

CHAPTER 4

STARDUST MEMORIES

QUANTITATIVELY, dust refers to solid particles with diameters of less than 500 micrometers. A micrometer, also known as a micron, is a millionth of a meter, or 0.000039 of an inch. The eye of a needle is 750 microns wide, enough to get some camel dust through. The diameter of the period that ends this sentence is about 450 microns—it would make a nice little piece of dust if it could be liberated from the prison of this page.

But despite its physical insignificance, dust has outsized negative connotations. It is an avatar of the unclean. It is the ensign of entropy, of buildings destroyed or neglected, matter without purpose. Dust is what gathers on books that are not read, on cabinets and shelves. It is the figure of the fragment, of division and disintegration, of the unswept, the unloved, the overlooked, and the discarded. It is what Tad Allagash snorts in his comical attempt to hoover any remaining traces of "Bolivian marching powder"—cocaine—in Jay McInerney's 1980s novel *Bright Lights, Big City.* To be clear, a Buddhist saying goes, you must wipe the dust from the mirror of your mind.

Much of household dust is keratin, the main protein of skin. Human skin sheds continuously. You lose about 1.5 grams of it per day. The detritus does not go entirely to waste. First, it is decomposed by fungi, who like it moist. Human flakes predigested by fungi are a staple for *Dermatophagoides farinae, Dermatophagoides pteronyssinus,* and *Euroglyphus maynei.* These mite species excrete protease enzymes that linger in mattresses and furniture—until the comparatively colossal rear end of a person sits down, at which time they are released in sometimes invisible puffs. The allergenic dust is a main reason for indoor sneezing, itching, and irritated eyes.

But dust ain't all bad. Illuminated by slanting rays, it becomes pixie dust. Like a wink of the Lucretian real behind all mythology, it becomes a princess kissed by the sun. Thus, this same humble substance that piles up on bookshelves, that darkens corners and lies flat on surfaces in an apparent effort to disguise its pitiable dishevelment (and why shouldn't it, after all, for what is dust but a debased form of dirt?), this asthmatic antithesis of the grandeur of existence, this figure of the unwanted so unremarkable it often escapes even negative attention, this forerunner of filth on the clichéd white glove of the fastidious housecleaner, this accoutrement of the homeless, this essence of crumbling, sometimes swirls and reflects, like glitter happy to be free of the mirror of our minds.

A dust grain can be a world. In Dr. Seuss's *Horton Hears a Who!*, a perspicacious elephant listens to a talking speck that turns out to be a tiny world on a grain of dust, one city of which, Whoville, contains a bustling community of humanoid Whos. I read the Seuss book to my son. It was a crash course in imaginary microbiology and crypto-anthropology, whose lesson was that within what we overlook are sometimes rich and unsuspected worlds.

When I was eighteen, in 1978, I would sometimes smoke angel dust in the bleachers with the black kids at lunch. This was in L.A. My fellow students bought it sprinkled on parsley. This is a dangerous drug that can cause very bad reactions. I was lucky that, for me, it just led to pleasant numb feelings under the California sun.

My father, Carl, was filming the *Cosmos* series for television at the time, and I was living with him and his second wife, sharing a room with their son, my seven-year-old half-brother Nick, now a successful science fiction writer. It was the first time I had lived with my father since my parents divorced when I was five. My dad let me hang out in his trailer, which had belonged to Marlon Brando, on the lot at KCET in L.A., where I would listen to punk rock on FM radio. I was from the East Coast, out of place, and had no friends to speak of. There was one girl I had my eye on, but that was about it. Blair High School was ethnically partitioned, one-quarter Hispanic kids, one-quarter black kids, one-quarter Asian kids who gravitated toward studies and chess club, and one-quarter spoiled white kids. In gym I had the chance to take modern jazz dance. There was only one other boy in the class. He was tall and effeminate, and got into a fight with a very dark girl half his size but twice as fast. She whipped off her two earrings, he whipped off his one, and they started flailing away at each other. This led to him getting kicked out. The white teacher asked me to stay after and wanted to know what I thought of the situation. I asked her if she was asking me because I was the only guy, and she said yes. "But you're the only white girl in the class," I said.

The next day after class, I was jumped by several boys on the steps of the little building past the pool. Phyllis, one of the black girls who knew me from the bleachers, was by the tennis courts and yelled, "Just run, white boy, as fast as you can to the principal's office. Don't look back." I did and identified the kids. For a few days, I stayed away from school, accepting my stepmother's painkillers, which she had for her back pain and which allowed me to forget about the fact that I'd have to return to school once the bruises and swelling on my face died down. When I returned to school, my tough black friend "Blue" informed me the kids I'd identified got kicked out and that there was a contract out on me. I did not take this seriously, considering it movie bluster inspired by *The Godfather.*

Aside from Blue, my only other friend at the time was Tim, a white boy who was gay and had an older boyfriend who was a professional therapist. One day, Tim and I were in my room listening to my David Bowie albums, whose songs like "Starman" and "Life on Mars" I half-hoped to share with my dad (not so much "Space Oddity," I explained, which was too popular). My half-brother, Nick, was playing superhero and jumped from the arm of the couch toward Tim, who stepped away. When he hit the floor, my stepmother had enough. She insisted my father speak with me. Soon thereafter, to my smug amusement, he lectured me, giving a rationalist discourse on natural selection as the basis for a normal heterosexual relationship. He underscored the evolutionary uselessness of Eros diverted from its natural goal. Instead of the Old Testament, he was appealing to Charles Darwin to make the same point.

I was glad he was paying attention to me, even if it was negative attention. The stories of neutron stars a spoonful of which weighed tons, of black hole portals to different places and times, and of higher-dimension objects moving through lower-dimension worlds that he'd told me as a boy he was now sharing with the nameless masses. Why, my unconscious must have wondered, was he spreading his message about contacting life in outer space to millions when I, his firstborn, had been neglected? Was I mere human dust, disposable detritus taken root in the fertile medium of my microbiologist mother, only to be forever after confined to the plutonic outskirts of his emotional universe? At least in Adriatic Italian dialect, as I would learn two years later from my college girlfriend, whose own undivorced parents were bhang-smoking partiers on Cape Cod who welcomed me into their home, there was a name for the dust you swept up—*mundessa.* Ah, to be embraced, named, recognized, even if only momentarily, on the way to the eternal dustbin.

Dust's relegation to a trope of dirt and insignificance, entropy and loss, is belied by its astronomical grandeur. And few, if any, have been more eloquent on this grandeur than my father, whose own bones now gather the stuff in sites next to his parents, moved from Florida to accompany him in Lakeview Cemetery in Ithaca, New York. From Martian dust storms as part of the genesis of his thoughts on how dust and smoke could affect life on Earth, to his study of Triton's streaks as windblown dust, to his "pale blue dot" speech, perhaps his most beautiful, in which he calls Earth seen from space "a mote of dust," my father took dust deadly seriously. And he led others to do so. He drew the world's attention to the role of particulate matter in blocking out light. Extrapolating from Martian dust storms, he tirelessly warned of a nuclear winter, the unsuspected result of smoke and dust rising in the aftermath of a superpower confrontation. The darkening blanket would, he suggested, shut down agriculture on a global scale. His advocacy and interaction with Soviet scientists played a role in ending the Cold War—which, ironically, was inaugurated by the Soviet launch of the Sputnik satellite that set off the space race and the attendant massive funding that made his spectacular career in space science possible.

My father was deeply moved by the tiny blue speck the Earth became when NASA, partly because of his advocacy, directed Voyager 1 to turn back in 1990 and photograph our planet. At the edge of our solar system, Voyager 1 relayed an image in which Earth was less than a single pixel (0.12 pixel) in size. That was us—a speck. My father's point was that every historical event, every sinner and saint, every loved person—and Voyager started relaying images on Valentine's Day—lived out their lives on this turquoise dot. Some 3.7 billion miles from its source, our planet looks like "a mote of dust suspended in a sunbeam."

Although he died in 1996, my father would also have liked another image, taken May 8, 2003, by the Mars Orbiter Camera of the Mars Global Surveyor, when Earth and Jupiter were aligned with Mars. Earth was about eighty-six million miles away and Jupiter almost six hundred million miles away. Because Earth is closer to the Sun than Mars, Earth appears half-illuminated, exhibiting a phase, like the Moon: a crescent Earth. Blown up, the image shows our Moon, as well as Callisto, Ganymede, and Europa, three of the moons of Jupiter and the inspiration for characters in serial adventure stories my father used to tell me, my brother Jeremy, and our friend David Grinspoon (now a professional astronomer) when we were children.

When my father was a child, his mother took him to the Brooklyn Public Library and introduced him to the librarian so he could get an answer to his question, "What are the stars?" After struggling with a boyhood facial tic, he overcame his shyness to ask the question, and a few moments later the librarian, saying she had just the thing, came back with a book on movie stars. Later, he became a television star. In *Cosmos*, he liked to talk about how we are "star-stuff."

We are not exactly star*dust.* The scientific definition of dust does not distinguish between mite droppings and pulverized diamonds streaming outside a cosmically careening spaceship. But dust must be solid, and stars are gas. Our Milky Way galaxy, with some four hundred billion stars, has a lot of dust. There is cosmic dust—space particles that may be only a few molecules in size—that takes different names depending on its location: intergalactic dust exists between the galaxies; circumplanetary dust around planets, such as in the icy rings of Jupiter and Saturn; interplanetary dust between planets; and interstellar dust between the stars themselves. Solid dust forms only about 1 percent of the interstellar medium, with hydrogen and helium gas forming most of the rest of it. Hydrogen, as H_2O, is the most common element in our bodies, as well as in stars, which turn it into helium in their core.

We are star-stuff also in the sense that not only hydrogen gas but other elements that come into being inside stars are distributed when they explode. A normal-sized star produces pressures and temperatures that turn hydrogen into heavier helium, but a supernova with twenty to one hundred times the mass of our Sun transmutes elements in layers like an onion via nuclear reactions in a natural alchemical process called nucleosynthesis. These elements are recycled when new stars, and planets, form. Our Earth formed with the other planets and the Sun from a rotating disk of ice, gas, and dust. Near the center of the protosolar nebula, intense pressures and temperatures vaporized debris, sending lighter materials away from the aggregating center and leaving grains of rock and dust consisting of heavier elements such as iron, silicon, and carbon to form the inner planets Mercury, Venus, Earth, and Mars, as well as their moons. Sly and the Family Stone were on the same cosmic track as my father when they sang "everybody is a star."

It would be interesting to deconstruct dust, but dust, as the fragmentary end-state of solid matter, is already deconstructed. After the recently contested Iranian elections, in which the Interior Ministry declared Mahmoud Ahmadinejad the winner by a "landslide," he tagged protesters as "dust and pebbles" who within "the transparent river of the Iranian nation" would find no place to "shine." The Bible, presciently anticipating modern knowledge of ecological recycling, advises us not to set store on things of this world with its tantalizingly transient treasures, as the matter of our bodies moves from earth to earth, ashes to ashes, dust to dust. But dust deserves its reconstructive due. Here astrobiology helps. It's possible that some dust in space harbors life. Life, as bacteria, can be extremely hardy. Bacteria live in solid rock on this planet a mile beneath Earth's crust. Bacterial spores are resistant to desiccation and radiation, and bacteria show far more metabolic diversity than all animals and plants combined. It is also possible that solar winds can distribute *Horton Hears a Who!*–sized dust grains containing bacteria across the universe, although they might not survive the cosmic rays. The Greek sage Anaxagoras gave this idea of a cosmos sprinkled with universe-traversing life its name, *panspermia,* from the Greek for "all seeds." The Swedish naturalist Svante Arrhenius liked the idea. So did Francis Crick, codiscoverer in 1953 of the structure of the DNA molecule (which discovery, it has recently been reported, was accompanied by the ingestion of low levels of LSD). When my father positively reviewed Crick's version of panspermia, which Crick detailed in his book *Life Itself,* I protested, "How does moving the problem of life's origin into outer space explain anything? It just transports the location of the problem." He conceded I might be right, but his heart didn't seem in it. In any case, Earth itself is in space, so even if life evolved here, it evolved in space.

Panspermia has the power to reinvigorate our notion of dust from a figure of neglect and unimportance to one of essential substance. Working with Sir Fred Hoyle (who coined the term *big bang*), the astronomer Chandra Wickramasinghe found decades ago that the infrared signatures of dust seen in all directions by peering astronomers matches the wavelengths of dried bacteria, suggesting that the cosmos might be full of bacterial dust. More recently, with his colleagues at the Cardiff Centre for Astrobiology, Wickramasinghe calculated that radioactive material such as aluminium-26, injected into our solar system during its formation from shock waves emanating from a nearby supernova explosion, heated the frozen centers of comets to produce subsurface oceans. Provided a liquid medium, bacterial dust may thus have come to life inside these hurling comets. Rocky on the outside, watery on the inside, they invert the composition of playground bullies' rock-filled snowballs. These astral bodies may be legion among the hundreds of billions of comets flying about just in our own solar system. Photos such as those of comet Tempel 1 taken by the Deep Impact probe in 2005 show evidence of ice core melting, which might happen as comets approach stars or burn through atmospheres. So maybe bacterial dust can stow away, sometimes coming to life inside spherical lakes, where they live in the dark, inside whizzing comets, for perhaps millions of years. And if one of the comets hits—voilà! They have light and the opportunity to evolve photosynthesis. Within a starry firmament filled with particles, a cosmos of violence and transformation, everything, even germs careening through space, may someday get the opportunity to shine.

CHAPTER 5

A QUICK HISTORY OF SEX

IN HIS *Letters to His Son on the Art of Becoming a Man of the World and a Gentleman,* Lord Chesterfield, the eighteenth-century British statesman and man of letters, offered the following concise account of sex: "The expense is damnable, the pleasure momentary and the position ludicrous."

Despite the droll nature of his quip, Chesterfield's observation highlights some deep truths about our status as living, breeding beings on this planet. The damnable expense—which in Chesterfield's case doubtless refers to the money and time spent in wooing, dating, and engaging in matrimony—theoretically applies to all sexually reproductive organisms.

Considering that some organisms can simply clone themselves—a well-fed amoeba grows and splits to produce two new amoebas—what is the point of making reproduction dependent on an intricate set of shenanigans with a member of the opposite sex but the same species?

When evolution can take one organism and create two, why make matters more difficult for itself by developing a method of reproduction that requires two organisms to make one? This added expense, in Chesterfield's terms, has led to a bevy of evolutionary theories on why sex exists.

So, too, the pleasure Chesterfield dismisses as momentary, and the position he describes as ludicrous, may have their origins in evolutionary theory.

. . .

WHEN WE ASK THE QUESTION "why sex?" the answer comes thundering back: "to reproduce." But that just begs another question: "Why reproduce?"

Sex and reproduction are not necessarily connected, even though they are strongly linked in us as well as in most plants and animals. In biological terms, sexual reproduction can be defined as the formation of new individuals from the genes of at least two different sources (i.e., parents). Simple reproduction, by contrast, is an increase in the number of individuals—they don't necessarily have to have any new genes.

Bacteria have been exchanging genes, without strictly needing to do so to reproduce, long before the evolution of plants, animals, fungi, or even amoebas. In some cases, one "parent" in an act of bacterial sexuality is not even alive; it's simply a raw gene—a DNA molecule in solution. This phenomenon was first demonstrated by the British medical officer Frederick Griffin in 1928 and called the "transforming principle." Griffin found that even dead bacteria of one strain could pass on their genetic material to live bacteria of another strain, thus "transforming" their offspring into the strain of the dead bacteria. It was later discovered that the "transformation" was actually caused by the living bacteria absorbing the DNA of the dead bacteria and using it to replicate.

We now know that viral DNA, genetic elements called plasmids, and whole bacteria with an entire set of genes may also serve as "parents" in bacterial sex. But no additional individual need be produced: the result of bacterial sex is not duplication but the same bacterium with a new set of genes. Then, when it actually does reproduce, it has the abilities or traits conferred by the new genes.

If we had sex like this, it would be like going swimming with brown eyes, picking up a gene for green eyes in the pool, and then passing that trait on to your children. Although it evolved on Earth billions of years ago, this kind of nonreproductive sex has lately been tapped en masse by the biotech industry. But the original biotechnology innovation was pioneered by bacteria.

There is tantalizing evidence to suggest that this primitive form of sex evolved in the billions of years before the Earth was enveloped in a protective ozone layer. This thin layer of O_3 is thought to have appeared only one to two billion years ago as a by-product of the metabolic process of green photosynthesizing cyanobacteria. The ozone layer around the Earth today shields life from the vast majority of damaging ultraviolet radiation.

However, bombard modern bacteria with levels of ultraviolet radiation similar to those before the ozone layer formed, and they disperse bits of naked and protein-coated DNA and RNA into the surrounding medium. This would have resulted in a veritable orgy of bacterial recombination in broad daylight.

One intriguing prospect is that strands of DNA damaged by ultraviolet radiation may have originally been fixed when primitive organisms shed loose genetic material and that this repair process was appropriated for recombination. But the sex that humans engage in to reproduce is decidedly different from that practiced by bacteria. In fact, the biggest distinction between life-forms on Earth is not between plants and animals but between prokaryotes (bacteria) and eukaryotes (all other organisms).

Bacteria have no nuclei and no true chromosomes, whereas eukaryotes do. Eukaryotes range from single-celled amoebas and *Paramecia* to trees, fish, and humans. Whereas bacteria can receive anything from a single gene floating loose in the surrounding medium to all of another individual bacterium's genes, the reproductive sex of cells with nuclei involves reception of half the genes from each of two parents. In truth, the search for the origins of our kind of sex must be sought in a study of these primitive eukaryotes.

CYCLES OF HUNGRY DOUBLING and genetic exchange—the origin of reproductive sex—are thought to have arisen at least three times in the history of life: in the swimming single cells called choanomastigotes thought to be ancestral to animals; in the swimming water mold cells called chytrids ancestral to fungi; and in the swimming populations of green algae ancestral to plants.

Once sexual reproduction arose, it went through many fascinating permutations and variations. Originally in ancestral microbes, gametes, or sex cells, were of the same size. The divergence into a relatively larger, more stationary gamete (the egg) and smaller, more numerous, faster-swimming ones (the sperm) set the stage for the separation into the bodies that housed them, that is, male and female animal bodies.

The original requirement for eggs to be fertilized in an aquatic habitat gave way to the rise of the amniote egg, which provided a protective environment in which the young could develop without being directly in the water. Whereas fish spawn their sperm and eggs directly in the water, selection naturally rewarded males who could deliver their sperm to eggs before they left the female body.

Penises evolved as an efficient sperm delivery technique. Geoff Parker, an evolutionary biologist at the University of Liverpool, put the idea of sperm competition, a form of natural selection at the level of the gamete, forward in 1970. Other things being equal, the competition among sperm of two or more males for the fertilization of an ovum rewards males with more sperm, larger penises, or sperm that displaces previously deposited sperm or has qualities such as extreme stickiness which forms a barrier, or plug, preventing subsequent suitors from delivering their genetic goods.

Banana slugs and Dungeness crabs, as well as many species of insects and rodents, are known to have sperm plugs. The damselfly uses his complex penis to scrub females free of rival sperm, that his own may enter her reproductive tract.

Nor are humans impervious to this form of rivalry at the level of sex organs and secretions: Experiments show that men who suspect their partners of cheating produce far more sperm per ejaculation than males persuaded of their partner's fidelity. Thus it is that jealousy, physiologically at least, is an aphrodisiac.

SCIENTISTS DISTINGUISH between the *origin* and the *maintenance* of sexuality, which is to say, why it exists despite, as Chesterfield put it, the "damnable expense." Ideas to explain the origins of sex—such as bacteria taking fragments of each other's genes on an ultraviolet-saturated primitive Earth—are not that easy to test experimentally. Scientists have thus tended to focus more on *why* sex exists.

Scientists believe that, although sexual reproduction is more complicated than simple cell division or cloning, it confers genetic benefits on the many species of plants and animals that engage in it. By requiring the fusion of genes in each new individual, sex provides an opportunity to pool advantageous traits and mutations. Additionally, when deleterious mutations come together in unfit individuals, those individuals die, thus ridding the population of negative traits.

There is some observational and experimental evidence to support these theses. In 1973 Leigh van Valen, an evolutionary biologist at the University of Chicago, put forward the "Red Queen" hypothesis, the forerunner of one of the most widely accepted theories of why sex exists. Van Valen's theory discusses how evolutionary change begets further change by altering the environment at large. The hypothesis is named after the queen in Lewis Carroll's *Through the Looking-Glass,* who says, "It takes all the running you can do, to keep in the same place."

Another evolutionary biologist, Graham Bell of McGill University in Montreal, explored the same principle. By shuffling their genes with each offspring, sexually reproducing organisms are better at eluding parasites. Evidence for this theory comes from experiments with minnows—small freshwater fish—in Mexico. Researchers separated asexually cloning and sexually reproducing minnows into different ponds. They then counted black spots caused by a parasite and found that the asexual minnows were far more prone to disease than their genetically more varied, sexed relatives. By shuffling their genes each and every time they reproduce, sexual reproducers appear to be better able to elude the evolving bevy of potential inner assailants.

Hybrid vigor—the noted superior hardiness of organisms with genetically distinct parents—is also associated with beauty, at least in humans. The chances of facial and body symmetry that we find beautiful increase when separate genetic stocks are brought together. This phenomenon can also be seen from the inverse; inbred fruit flies have less symmetry. Cheetahs, having little genetic diversity among remaining populations, tend to have asymmetrical facial bones.

Clear, disease-free skin is a key trait in human estimations of beauty. Thus, when lovers are attracted to each other, they may be unconsciously estimating each other's freedom from, or ability to ward off, potentially damaging parasites, which have a harder job attacking the moving target of obligatory genetic recombinants.

Yet, while much ink has been spilled over what "maintains" sex (why it continues to exist in sexually reproducing organisms), it's not as though we humans, and other plants, fungi, and animals, can just opt out of sexual reproduction. In multicellular beings, sexual reproduction is the only way a new being can form.

Sexual fusion of eggs and sperm is required to start the embryo, and, except for certain exceptions such as the green algae and the artificially selected triploid banana, all plants produce embryos, as do all animals. On paper, it is easy enough to say we would prefer to eliminate the "damnable expense" and just be cloned. But, as technology stands, even the wealthiest misandrist (or hater of men) cannot access the technology necessary to bypass males if she wishes to reproduce.

A similar situation appears to be the case in the animals officially classified as "asexual." Careful study reveals that these organisms—such as the all-female populations of whiptail lizards that inhabit the southwestern United States—still undergo chromosome sex processes at the cellular level. In the case of the female whiptails, they also mount each other and produce more eggs when another female has mounted them. Thus asexual animals appear to be evolutionary rarities that have devolved from a sexually reproducing state, rather than truly "asexual" beings.

To understand the depths of sexuality, we don't just need to study animals and microbes, for the roots of sexual reproduction lie in events that are many hundreds of millions of years old. The Red Queen, and other theories explaining why sex is genetically worth the trouble, assumes that most species of plants and animals could simply lose the trait of sexual reproduction if it were adaptive. But this may simply not be the case.

We may wish we could fly because it would be more "adaptive," but the physiological option is simply not open to us. Sexual reproduction may be so old that most species are unable to lose it. And those that do may become vulnerable to rapidly spreading parasites, and therefore removed from the gene pool. Simon Robson, a biologist at James Cook University in Queensland, says of zoologists who theorize on the maintenance of sex without regard for its nonanimal history: "They're working from a data set two billion years out of date."

In fact, in the 1940s the Harvard University biologist Lemuel Roscoe Cleveland solved the mystery of the origin of fertilization and how cells with one set of chromosomes (haploid cells) came to merge with each other and develop two sets of chromosomes (becoming diploid). Cleveland noticed that in some cases, hungry amoebalike cells would incompletely devour each other, merging cell membranes and cytoplasm and pooling their chromosomes in a single nucleus. He reasoned that the first fertilization was the result not of a sexual but a starvational urge to merge.

This theory is borne out by evidence from other quarters. Some cells today reproduce asexually . . . at least until they are faced with starvation, such as from lack of nitrogen in their environment. In this case, they fuse together to make a diploid. Primordial sexual fertilization, which Cleveland witnessed among microbes called hypermastigotes, may have had a most unromantic genesis. Cleveland also carefully chronicled and photographed the necessary separation from the state of doubleness, which he called "relief from diploidy." We undergo such relief when we produce haploid sperm cells or ova from diploid body cells; aside from these sex cells, the rest of the cells in our body remain in a diploid state, with forty-six chromosomes, twenty-three each from mum and dad.

THE KINKIEST HUMANS are utterly unimaginative compared with the sexual variety on display in the animal world. African bedbug males, for example, routinely pierce females through any part of their carapace in order to impregnate them.

Natural selection is amoral. Which dysfunctional human family is comparable to the behavior of *Adactylidium,* the nasal mites whose mothers give rise to 99 percent females? Inside the mite's womb, one unborn male inseminates all his unborn sisters, then dies. The incestuously impregnated females grow, eating their mother's body from the inside out to provide themselves with nourishment. They are born pregnant, beginning the cycle again.

Although we are accustomed to regarding serial monogamy as normal, our hominid ancestors may well have regarded us as deviants. Robert L. Smith, an entomologist at the University of Arizona and a leading theorist in sperm competition, suggests that human ancestors *Homo erectus* were more promiscuous. The mating systems of our closest relatives, the great apes—gorillas, orangutans, and chimpanzees—differ. Gorillas are sexually possessive and live in "harems" dominated by older silverbacks, and a single male dominates females and males. Orangutans are loners who occasionally come together to copulate. Chimpanzees, however, are relatively promiscuous. A female chimp in heat will develop a red and swollen pudendum and copulate with virtually every sexually active male in her troop except for her own sons.

In troops of chimpanzees, if couples go on "safari"—or disappear to be alone together in the woods—as often as not they are beaten upon their return. The group considers elopement behavior, normal for human lovers, aberrant. Whereas big gorillas have erect penises that measure just 2.5 centimeters, promiscuous chimpanzees produce more sperm than any other great ape. Humans are the second most copious sperm producers among our genetically closest primates, suggesting to sperm competition theorists that our ancestors were more promiscuous. Today we have a "mixed" mating system that resembles both the "free-loving" chimpanzee and the "patriarchal" gorilla styles. We are at the crux of two evolutionary approaches to coupling, with neither being any more "natural" than the other.

The unusual vulnerability of human infants has evolutionarily favored females who lost their putative ancestral period of visible ovulation—the estrus with the accompanying reddening that attracts chimps. Such females were more likely to interest sex-obsessed males throughout the month, increasing the chances that they would stay around for child care. Some suggest that culture itself, from poetry to painting to music, is a by-product of ancestral human females attempting to find outward expressions, and exaggerations, of male genetic fitness.

Human intellect in some sense may thus be the mental equivalent of the peacock's tail feathers: sexual displays aimed at attracting a mate. Whatever the true story of the origin of sex, the quirks and oddities of sexual behavior show that normality is a fleeting commodity. Each new sexual species renders its ancestors obsolete as it incorporates new behaviors and new pools of mutating, recombining genes.

THE EVOLUTION OF MATING SYSTEMS knows few bounds. Among the plethora of life-forms on Earth, sex is manifested in a staggering variety of ways.

Many insects produce spermatophores: packages of sperm inserted into females or given to the female for self-insertion. Male octopuses use one arm to deposit sperm. Male anglerfish are diminutive creatures that attach themselves to the genitalia of females. Some wasps mistake orchids for females, pollinating the flowers instead of inseminating their own kind.

Then there's the laughing hyena, which hunts in all-female packs. Hormonally masculinized, their clitorises are longer than the penises of the males. They give birth through their urethra, and the cub's passage through the U-shaped birth canal usually kills the mother.

Even when it comes to the division of organisms into discrete genders, things aren't black and white. Only 5 to 7 percent of plants have males and females separated into distinct organisms, in the way that most animals are. Twenty to 30 percent of plants have male and female flowers on the same organism. The majority of flowering plants, such as tulips, have both sexes in the same flower, and they mature at the same time.

People worry about tiny differences within our own human mating system—but there really is no such thing as "normal" in sexual evolution. Take the great grey slug, *Limax maximus,* whose hermaphroditic midair mating antics first attracted attention over a century ago. Meeting up on a branch or bracket fungus, the lovers kick off with a bit of tentacle foreplay. Then they close in on each other, circling in a mouth-to-tail dance that lasts up to two and a half hours.

Then, at the point of utmost passion, each slug "bungee jumps" from the tree, stopping itself short on a 38- to 46-centimeter line made of its own mucus. Swinging in midair, each slug unsheaths its penis, up to 10 centimeters long, and inserts it in the appropriate bisexual organ. After exchanging sperm, they either climb up the mucus cable whence they came or drop to the ground from the sheer orgasmic exhaustion the French call *la petite morte* . . . the little death.

The exchanging of genes in sexually reproducing species has a long and titillating history whose variations transcend the imagination of the most dedicated pornographer. From DNA repair mechanisms in bacteria to flower images sent over the Internet, sex in its great diversity can be expected to persist beyond the demise of the human species. Or, alternatively, be involved in our evolution into new species.

CHAPTER 6

WHO IS I?

AT THE END of the year last, in a party diverse with ethnicity and artistry, not to mention anarchists, a question was asked of me by none other than myself. Yet, as it was done in company, I credit the question as much to my companions.

You see, after I described some of my political views, mentioning the strange question of the status of the Federal Reserve as a private corporation, as well as some of the scientific anomalies surrounding the events of 9/11, I was told that my views pretty much matched those of members of the Tea Party. Now I knew I was against the neocons, but I had no idea that, according to a helpful anarchist, that made me a fledgling member of the Tea Party.

I don't watch TV news and now find most of the alternative media as noxious as the mainstream kind. Which doesn't leave me much of an informational safety net except for the Internet, which, as we know, is full of holes. Yet that's where I, *insofar as I am an I,* swim in a roiling sea of glorified gossip and the occasional fresh tidbit of gleaming truthlike debris.

Of course that's also where *Wild River Review* (where this chapter was first published) is, so it is at the very least convenient for this essay. But it also means I was inured from my status as a putative Tea Party nutcase (not to be confused with nutjob, nutbag, or nutbar). I had hardly known that I had metamorphosed in my sleep. But in case there was any doubt, an ex seconded the motion on Facebook, wondering since when had I developed an affinity for the extreme right wing.

Well, I never.

When I asked the anarchist if he thought Julian Assange of WikiLeaks fame was a double agent (because his revelations had apparently been vetted by Israel, and because such an operation could be used as an excuse to shut down the Internet in the name of security), he assured me that he (Assange) could be a *triple* agent, which I found nicely to my liking. Assange could be fooling both sides.

Not to be outdone, I spent the rest of the party spicing up my conversation with the perplexing notion of the quadruple agent, a concept around which it is indeed hard to wrap one's mind. The anarchist added that Sarah Palin, who later terrorized the country with her gunsight campaign graphics, was probably, at least financially, a creation of the neocons.

Right wing. Left wing. What?

Needless to say, I was confused. But the confusion did me good, as it allowed me to muse on some of the rarefied niceties of that perplexing morass of abstract marble from which we shape ourselves into selves. I speak of identity. I can answer where I am (on the third rock from the Sun, in the outskirts of the Milky Way), what I am (proteins and genes and bones and whatnot made from atoms common in the universe), how I am (okay), when I am (twenty-first century, etc.), and maybe even why I am (more on that later) more easily than I can tell you *who* I am.

Apparently I am not alone in this perplexing dilemma. Although it will do no justice to paraphrase the great popularizer of mysticism and expositor of world religions Alan Watts, who devoted a whole book to the subject (*The Book: On the Taboo against Knowing Who You Are*), Watts gave an elegantly Hindu answer to the question: you are the universe playing a game of hide-and-seek with itself.

Basically it boils down to this: Because the universe is eternal, which can get boring, it likes to pretend that it is divided into individual parts. Some of these parts not only die, they *know* they die. This realization may get those parts worked up, but it also keeps them from being bored.

Watts pointed out that the universe doesn't like to show off in this regard, but rather to "show on"—as a character would show up in the pages of a book, or an actor upon the world's rotund stage. Later in his life Watts distanced himself somewhat from this Hindu metaphysics, but *The Book* still stands as a brilliant testimony to one of the simplest and most enduringly convincing ideas of cross-cultural religion: We are bits of the all, the cosmos engaged in a grand game of self-play.

Watts here and elsewhere espoused a doctrine of realistic reincarnation. He looked at his red-headed grandchildren and saw himself. He opined that all organisms "think they're human."

And he spoke of the need to escape from the illusion of the "skin-encapsulated ego" in order to recognize the connection in each to the infinite. Turning a noun into a verb à la Heidegger, he said that when a baby is born the universe "I's" itself. It is the same cosmos I-ing itself in a myriad of forms. We may die, but the great game goes on. New beings are born, but they make the same discoveries. They, in a sense, "R" us.

(And I think it's weird how in English we ask how "are" you—as if in addressing another in the second person we are somehow secretly acknowledging their multiplicity. Shouldn't it be, how is you? Or maybe, what you be?)

Samuel Butler, novelist and neglected philosopher, is an interesting case in the exploration of multiple identity. Butler wrote of walking down the street and noticing that every person reminded him of someone else. So this guy might look most like the Earl of Sandwich, another like Jesus, a gal like Queen Victoria. We're familiar with this syndrome from its extreme form in the asylum. Butler just entertained a light version, Watts's naturalistic reincarnation applied to the other rather than the self, reincarnation right out on the street.

Butler experienced other disruptions of the self. His *Erewhon: Across the Range*—a utopia that combined New Zealand and northern Italy, the Maori and the English into a satiric fictional blend—was originally thought by the public to have been written by Sir Thomas More. It sold briskly until its real author was divined, and sales dropped. Still, for Butler, the experience of being taken for another was not entirely displeasing.

Butler had already played at identity by arguing with himself in the op-ed pages of *the Press* in Christchurch, New Zealand (where on February 22, 2011, there was an earthquake measuring 6.3 on the Richter scale). One of Butler's avatars ("Cellarius") took the position that machines (of which the telegraph and train were the most advanced examples of the time, 1865) were taking over the planet, whereas another, anonymous author (writing a piece called Lucubratio Ebria, Latin for "drunken nightwork") scathingly disagreed with Cellarius, pointing out that devices like umbrellas were extensions of our skin and that a train is only a "seven-leagued foot that five hundred may own at once."

The author of fictions indulges in this same sane version of multiple personality syndrome; she is the outermost concentric personality who knows the true status of her characters while they, poor saps, have nary a clue, mostly, of the whimsy-driven coffee-drinking goddess controlling their fate.

Indeed, Socrates's dialogues in Plato's hands were arguably the earliest modern novels, as they allowed a panoply of distinct voices to transcend the limitations of isolated opinion in order to create a multipersonal philosophy beyond individual opinion. The irony here is that Plato, according to Friedrich Nietzsche in *The Birth of Tragedy out of the Spirit of Music* (great title, that), had originally planned to be a dramatist, that is, a tragic playwright.

Under the influence of Socrates's rationalism, however, Nietzsche supposedly burned all his plays as a mystical enterprise unbefitting the truth-telling agenda of the philosopher. But what the Socratic method sacrificed, according to Nietzsche, was the realization that the multipersonal realm of Attic tragedy was a sacred reenactment (a "showing on" in Watts's lingo) of the primordial drama of human separation from the cosmos that defines us.

Euripides, also under the influence of Socrates, started it by getting rid of the chorus. A central part of the essentially spiritual ancient Greek tragedies, the chorus was not meant to be taken literally as representing people but was, instead, a manifestation of the cosmic realization that we are all one, illusorily separated from one another like raindrops glinting in the sun as they fall, unaware that their source and destination is the current of an indivisible river. Getting rid of the chorus paved the way to melodrama and soap opera, to simple representations of daily life obscuring the tragic truth dramatized by the ancients, that our separateness is life's temporary illusion.

Butler's deconstructions of identity, his divisions of the would-be indivisibility of individuality, also took a biological turn: is it not arbitrary to identify death as occurring at the transition from maternal butterfly to eggs? Is not the transition from egg to caterpillar, or caterpillar to chrysalis, or, most spectacularly, from pupa to winged form, equally as striking?

He argued also that as infants, we are more like other infants than ourselves as octogenarians. The arguments were of a piece, with one another and with Watts's bombshell in *The Book*: "We" are not what we think we are, the stable identities conferred by pronouns like me, my, and mine.

Instead, in the words of the Vedic Sanskrit hymns, the *Rig Veda*, "thou art that": We are not just within but outside our skin, like waves connected to the whole ocean. Our true identity is (to use a Butler term) extracorporaneous: It is the universe itself, glittering forth galaxies, solar systems, planets, and beings.

MY COWBOY FRIEND Bill Huth, longtime owner of the Willow Springs Raceway in California, is a devout reader (and sometimes publisher) of old texts on evolution and spirituality. Huth argues that what we call "life" is eternal, evolving, a restless "thing" of ever-changing forms. This makes each of us eternal whether we know it or not, and generally we don't. Huth himself, now eighty-seven but once a tireless and quite accomplished conman, was accused multiple times in his earlier years of impersonating a preacher.

Once, he laughs, the police called his mother in L.A. to tell her.

"Oh no, he's not a preacher," she corrected them. "He's an evangelist."

The great literary scholar and surrealist Jorge Luis Borges was a master at revealing the subtle ways we are not who we think we are. In *Borges and I,* he writes of the "other one, the one called Borges" whom he knows "from the mail" and whose name he sees "on a list of professors or in a biographical dictionary . . . but I recognize myself less in his books than in many others or in the laborious strumming of a guitar. . . . Thus my life is a flight and I lose everything and everything belongs to oblivion, or to him. I do not know which of us has written this page."

In *The Other,* Borges depicts himself in Cambridge, Massachusetts, stopping to sit on a bench on the bank of the Charles River. Already on the bench is a well-dressed man who seems to him familiar. The older and the younger man have a refreshingly literary conversation, touching on, among other things, the use of the doppelgänger in the fiction of Edgar Allan Poe. At some point the older man realizes why the younger man looks so familiar; it is his younger self whom he, at first, did not recognize.

I had the exact opposite experience when I recognized myself, not on a bench as another body but as a voice from another time. My mother had picked up me and my girlfriend Natasha Myers, a professor of anthropology at York University, at the airport and insisted we listen to an essay of mine, part of an anthology that had just been released as an audiobook. The reader, Pamela Ward, was espousing views using my exact words but in prose that ran strongly contrary to some of my current opinions critical of critical theory, epistemological relativism, abstract jargon, and postmodern academic fashion.

Not only was she confidently using French philosophy to deconstruct the notion of discrete identity, I was the source of the voice and its mannered attack on the recent positions I held so dear. I shrank in my seat. Natasha, a frequent interlocutor with whom I had espoused my current opinions, was in the backseat, laughing.

The essay in question, "The Uncut Self," appeared in the anthology *Dazzle Gradually: Reflections on the Nature of Nature*, published by Chelsea Green. I had written a draft of it over twenty years ago for Fred Tauber of Boston University and his conference there, "Organism and the Origins of Self," on biology and philosophy. I was under the spell of Continental philosophy in a big way, quoting Michel Foucault, thinking with Jacques Derrida, and in general making the same sort of arguments that I now objected to when Natasha made them to me, accusing me of scientism, biological reductionism, and a naive belief in reality free of social constructions.

Dazzle Gradually comes from Emily Dickinson's line "The truth must dazzle gradually / or every man be blind." Well I was dazzled all right. And all would have been well had not my mother, the evolutionary biologist Lynn Margulis, coauthor of the essay and lately taken by it in audio form, insisted on playing it loudly for Natasha in the backseat.

Natasha could no more control the volume than I could, as my mother told us to "shhh" and listen, wondering what was so funny. My laughter was more subdued. I was being schooled by my younger self.

. . .

NATASHA AND I TRY not to make a habit of arguing, but when we do, our intellectual arguments seem to follow the script from an Ian McEwan novel, with me defending the "rational male" view of classic scientific objectivity and her assuming a more "generous" view (popular today in humanities departments) that highlights the role of culture and history in creating what we naïfs consider culture-free facts.

Part of the disorienting effect of the essay is that it begins midsentence:

> *full circle, not based on the rectilinear frame of reference of a painting, mirror, house, or book, and with neither "inside" nor "outside" but according to the single surface of a Moebius strip. This is not the classical Cartesian model of self, with a vital ensouled* res cogitans *surrounded by that predictable world of Newtonian mechanisms of the* res extensa*; it is closer to Maturana and Varela's conception of autopoiesis, a completely self-making, self-referring, tautologically delimited entity at the various levels of cell, organism, and cognition (Maturana and Varela 1973). It would be premature to accuse us therefore of a debilitating biomysticism, of pandering to deconstructive fashion, or, indeed, of fomenting an academic "lunacy" or "criminality" that merits ostracism from scientific society, smoothly sealed by peer review and by the standards of what Fleck calls a "thought collective" (Fleck 1979). Nor would it be timely to label and dismiss us as antirational or solipsist.*

What was "I" thinking? Where was "I" going with this? Well, apparently back to the beginning, because the essay ends: *Topologically the self has no homuncular inner self but comes* . . . thereby beginning the sentence with which the essay starts.

The effect of calm intellectual self-annihilation was complete. "I" had proved to "my" own satisfaction that there are no absolute borders around the self-referring, operationally closed but multiply constituted self. If God, as Meister Eckhart said, is a being whose center is everywhere and whose periphery is nowhere, so the self was not a skin-encapsulated ego separate from the rest of the universe.

I guess this also included separate from one's former self. I shrank from the auto-onslaught, but there was nowhere to go in the auto. The effect was worsened as Natasha alternately told my mother to turn the sound down and shrieked with delight in the backseat at my anticipatory self-destruction. The irony was that the self on the CD was deconstructing the older-me listening even as it embarrassed me with its jargon bomb of jejune enthusiasm. If the old young me was right, the new old me was wrong.

I thought of Nietzsche's comment about how juvenile his earlier writings seemed until he got older and considered that critique juvenile. Not only did Ward's reading of my essay embarrassingly reproduce Natasha's side of the arguments we have over "epistemological relativism" and "postmodern jargon," but my younger self argues *against* the notion of stable biological identity (the "rectilinear self" in that essay's jargon!). It was a one-two punch I gave to myself, and there was nothing much I could do but stay slumped in the front seat, pummeled in part by my own embarrassed laughter. Of course my mother still wanted to know what was so funny, as she thought the essay was "just great."

So, who are we?

I believe we are distributed identities, Möbius strips (okay, younger self, don't gloat) that turn back on ourselves to see that we are not the isolated simple identities we thought we were. We are members of families, tribes, nations, age groups, sexes, trades, and classes that may and are in complex conflict with one another. Navigating these multiple assemblages is an invitation either to contradiction or to denial; political coherency becomes impossible.

The multiple alliances go still farther. You are not just a political animal but a differentiated clone of nucleated cells, a collection of microbes. You are a lineal descendant of the first life, recycling a water-based chemistry full of hydrogen-rich compounds, like methane and sulfide, characteristic of the inner solar system four billion years ago at the time of life's origin, soon after the Sun turned on.

Atomically, you contain elements like carbon and oxygen, made not here but on the inside of distant stars that ultimately exploded. Your lineage escaped several serious mass extinctions, not including the global pollution crisis precipitated by the first water-using photosynthesizers that toxified the entire planet, but whose air you now breathe. Physically, you may only be a tick in time and a speck in space, but ultimately you are part of the evolving universe itself, much bigger than humanity and its current crop of madmen.

If you look on the Internet you will see that Julian Assange's last name is that of his stepfather, a theater director his mother married when Julian was one, and that several years later she married a musician who Julian himself says may have been part of the identity-destroying Santiniketan Park Association, a bizarre Australian cult run by Anne Hamilton-Byrne, who fed her children LSD, starved them, provided them with a rigid regimen of yoga and early rising, and made them repent for their sins as well as call her God. The children's hair was dyed platinum blond, the boys were given bowl cuts, some of them disappeared, their names and birthdates underwent capricious changes, and they often had multiple passports. Julian's mother was so intent on escaping her second husband that she took Julian and his half-brother into hiding, and by the time Julian was fourteen, they'd moved thirty times.

It has been suggested that the cult was an MK-ULTRA operation, a covert, illegal CIA human research program, run by the Office of Scientific Intelligence, and that WikiLeaks is a "limited hangout" in CIA jargon, that is, something that seems bad but is really good (for them).

I would like to suggest that the Department of Defense created a monster in helping form the Internet (in the late 1950s, partly to have a decentralized communications network that could survive a nuclear war) that they can never stuff back in their Pandoran box; that our complicity with the military industrial complex, reinforced whenever we drive down the street or use a MasterCard, seems to be under control of the paranoiac-conspiratorial-realistic "They"—some of whom really do fashion themselves to be in control—but that if we look farther into the future, complicity with THEM pales next to the power wielded over us by nature.

The biosphere has other plans for the Internet, a kind of global neuronal intelligence, that even the criminalocracy can't stop or control. Even billions of military dollars could not reproduce something as simple and lovely as Smith College's greenhouse across from Paradise Pond because that greenhouse is a very specific growth form, produced not just by humans but by plants in interaction with them for over 150 years.

We are complicitous with the forces we detest, but those forces are complicitous with still greater powers we may admire. The great heterobiographies of our protean selves have yet to be written. We are all triple agents now.

CHAPTER 7

OF WHALES AND ALIENS

The Search for Intelligent Life on Earth

HALF MY LITTLE LIFE AGO, under the influence of *P. cubensis*—aka psychedelic mushrooms—I, and two of my reprobate friends, found ourselves among a sea of tourists in Quincy Market. After overhearing a mini Sopranos-style imbiber declaiming loudly upon the niceties of female lace, frilly clothing, and all things that tied, we shambled on through the colorful commerce toward nature, or what was left of it down near the harbor.

We found ourselves noses to Plexiglas at the outdoor tank of the Boston Aquarium, attempting to make "Hoover," the great bull seal and for us the aquarium's main attraction, speak. Hoover was a character: After diving and holding his breath, he'd release spiral swirls of air bubbles like rustling aquatic theater curtains, building suspense for the performance just to come.

Then he would emerge and bellow such gems as "urgh-urgh-urgh hell-hell-Hell-HELLO How AH ya?" or "Guh-guh-guh-GUH-GUH-GUH-GEHT-outta-HEAH."

This time, however, although we would dearly have loved to have seen the best free show in town, the talking seal, despite our loud imprecations, did not respond, preferring apparently to wait for a larger audience or to slumber amid his substantial and slippery harem.

Our efforts did not go unnoticed, however. A drunk on his bench awoke from his slumber. "Hey-ya," he yelled, "get offa ThEAH!"

While Hoover got his Boston accent not from that drunk who slept on a bench near the tank but from the Swallows, the Maine couple who named him for his vacuumlike capacity to down fish—and then, when he grew too big, gave him to the Boston Aquarium—there was something enchantingly kindred about him, so much so that he received a human-style obituary from the *Boston Globe.*

With such intelligent mammals in the oceans that cover two-thirds of this watery orb we land animals have christened Earth, I wonder why, for the first time in twenty-four years, the International Whaling Commission has found it necessary to attempt to roll back the ban on commercial whale hunting. Even the hunters of whales realize they are tempting fate. *Moby-Dick* tells us of the Nantucket legends of the first indigenous harpooners, rowing toward the legless Leviathans of the deep. By Herman Melville's time, the industry had become both lucrative and romantic, attracting young men from across the continent to the urbane port of New Bedford, then a most cosmopolitan city itself coursing—with a questionable captain—through the greater ocean of space. Before setting off, Ishmael takes in a sermon for sailors delivered by an ex-harpooner priest. As he listens, the wood of the pulpit reminds him of a ship's prow:

> Then God spake unto the fish; and from the shuddering cold and blackness of the sea, the whale came breeching up towards the warm and pleasant sun, and all the delights of air and earth; and "vomited out Jonah upon the dry land"; when the word of the Lord came a second time; and Jonah, bruised and beaten—his ears, like two sea-shells, still multitudinously murmuring of the ocean—Jonah did the Almighty's bidding. And what was that, shipmates? To preach the Truth to the face of Falsehood! That was it!
>
> And in guiding the young mariners, Father Mapple, the ex-harpooner cleric, seems himself to lose his moorings: "He said no more, but slowly waving a benediction, covered his face with his hands, and so remained, kneeling, till all the people had departed, and he was left alone in the place."

Unlike our oceangoing legless cousins, we landlubbers are good at killing with our hands. Before their smiling visages became a fixture at SeaWorld and on the face of *Flipper* in the 1960s TV show, dolphins were disparaged as "herring hogs" for their tendency to rob fishermen's nets.

Sea mammals were hunted for lamp oil, for meat, and for "superior lubricant for precision timepieces." While alive they could be feared, but in the main they were treated as resources, not beings. A member of the "toothed whales" (Odontoceti), which also include narwhals (Melville's "sea unicorns," which have big tusklike teeth jutting out of their foreheads) and beluga whales, in the suborder Delphinidae, which also includes "killer whales" (or "orca"), and beaked and pilot whales, dolphins—largely because of the militarily funded, brilliantly creative, and somewhat unhinged researcher John Cunningham Lilly in the 1960s—came to captivate the human imagination.

I HAVE A VERY EARLY MEMORY of being with my father, Carl Sagan, who was trying to talk to a dolphin as another man walked about. It was probably Lilly.

My mother, Lynn Margulis, who would have been divorced from my father at the time, suspects I did meet Lilly, although not in the Virgin Islands at his main lab, but in Boston or Cambridge or Florida. She herself met Lilly once through my father and instantly thought Lilly was "clever and self-centered, much like Timothy Leary."

In his biography, William Poundstone writes of how my father, at a restaurant with Lilly, asked their pretty waitress out. Although she declined, she agreed to one of Lilly's crazy experiments, sharing a special flooded living quarters with a dolphin—who happened to be one of the five that played Flipper on the TV show.

Apparently Flipper had needs. (He was not alone; according to Princeton University's D. Graham Burnett, these "powerful sea mammals with fixed grins [that] now and again . . . rake, butt, and sodomize each other . . . have presented challenges to their keepers from the earliest days of captivity.") The waitress-cum-interspecies-experimental-subject found herself following a path of least resistance that included acquiescing to Flipper's relentless sexual advances, satisfying him with her hand. The dolphin was not, according to Poundstone, so lucky with my father, who in turn rebuffed the TV star.

Perhaps dolphins are not as smart, noble, or linguistic as we'd like to project. Dolphins, Burnett says, "though they can jump almost twenty feet in the air . . . very rarely sort out how easy it would be to roll over the top of an encircling trap." But is it not also true, as the recent British Petroleum Earth Day oil spill suggests, that we are also in a kind of trap, one not confined to a little net in the middle of the Atlantic Ocean but extending across all the continents and the seven seas, into the groundwater and the atmosphere, a trap that may spell the demise of ourselves as well as the dolphins and whales we cavalierly kill?

The cosmologist Stephen Hawking has recently blitzed the media with warnings not only that aliens probably exist but that when they find us they'll probably want to eat us. To me this also sounds like projection. Coming here for food is not like the harpooners setting sail to procure whale oil. Interstellar spaces are far greater, and the proteinaceous rewards far more meager. Distant aliens coming here for dinner makes about as much sense as flying a supersonic jet to Morocco for a garbanzo bean.

The physicist Michio Kaku counters Hawking's surmise that aliens may prove as destructive as Christopher Columbus and company when they wiped out Native American populations. If aliens arrived, Kaku opines, it might be more like the United States' experience during the Vietnam War, with the aliens wanting to get out ASAP.

To me, Hawking's Columbus and Kaku's "Vietnam" scenarios seem, to quote Friedrich Nietzsche, all too human. They relate our hypothetical meeting with aliens to other examples of human encounters within the history of our own dangerously self-absorbed species. Indeed, Hawking, bless his cosmic book-selling heart (in an interview on *Larry King Live,* he told King it was nice to see him after these ten years and closed by saying he hoped to see him again when his book came out), advised Larry that the invaders will "have a mouth opening because they will have to take in nutrition . . . and they will probably have legs because they will need to move around, and they will have eyes—but don't expect them to look like Marilyn Monroe." Whew—that *would* be scary—being eaten by an army of Marilyn Monroes from outer space!

Seriously, though, what is this obsession with hypothetical aliens when we are living among some of the most fascinating sentient beings in the universe and have only just begun to establish contact with *them*?

What does it mean to communicate with an "alien" when we barely have the first idea of how to understand the intricacies of rain forest plant communications, nuclei-trading mycelial networks, and ultraviolet light-detecting superorganisms of bees, let alone the conscious or unconscious minds of a humpback, great blue, or white whale—whose mathematical, philosophical, aesthetic, and perceptual abilities may, for all we know, far outstrip those of our greatest geniuses (whose intelligences we also don't judge by their skill in bloodshed)?

In 2008, at a conference in Basel, the ethnopharmacologist Dennis McKenna characterized psychedelic drugs as "molecular messages sent by Gaia," one of whose messages is that we—like monkeys excitedly trading glowing bits of a mysterious crashed starship found in a jungle—are not so smart. Indeed: Why should we worry about hypothetical interspecies communication (even Hawking says intelligent aliens are unlikely to exist within one hundred light-years or we would have detected them already) when we do not even understand the local "starship" of Earth's many species?

Burnett may be right that dolphins are not that smart. But we may not be that smart either. Nonetheless, if brain size has anything to do with it, some individual whales may be far smarter than individual human beings. Considering how much extra intelligence goes on beyond the level of conscious rational awareness—in our immune system, in our physiology, and in our intuitions—it's almost overwhelming to consider the capacity of a white or blue whale's unconscious mind. Despite their size (which you'd think would make them easy for our scientific sleuths to track)—the biggest ever in evolution, weighing two hundred tons each, their hearts as big as a car, their brains ten times more voluminous than ours—we do not even know where the blue whales go to breed.

If such whales with whom we share this oceanic planet remain deeply mysterious, intelligent aliens in our midst, the same may also be true of a far larger being, even closer to us. I speak of the planetary biosphere of which we humans seem to be minute parts, not unlike some of the cells of our own bodies that, if they are sentient, which some may well be, likely have zero conception of the coffee-sipping, car-driving wholes of which they are part.

Fossil and mass spectrometric evidence strongly suggests that Gaia—the visionary scientist James Lovelock's name for this systemic, cybernetic, intelligent-acting nexus of life-forms at Earth's surface taken as a physiological whole—regulates the chemistry and temperature of our planet's surface in the unconscious manner of a living being. Whales are mammals with whose sentience we can empathize, even if we can't understand them or they us. But, conscious or not, the living biosphere appears to be a far bigger fish, so to speak, one whose existence we've barely divined.

If we are to worry, we should worry about this complex beast of a planet with whose vast, mostly unconscious living intelligence we have been seriously meddling. Compared with the possible actions, which may soon be visited on us by this leviathan, the alien of which we are a metastatic part, the concerns voiced by our charismatic physicists are distracting at best, irresponsible at worst.

The media-grabbing headlines about being detected and eaten by distant aliens seem histrionically misplaced. The cosmological worry-warts are not providing a public service so much as displacing our local guilt over degrading and killing off whole species of our own very real and close relations. I'm not laughing, and I signed the petition to keep alive the ban on whale hunting. But it would be a rather fitting bit of cosmic irony if this giant intelligence, this greater leviathan of which we are part, this Brobdingnagaian body we are feeding on, this living surface of Earth whose physiological abilities still remain unknown and thus in a sense alien to the majority of people on Earth, itself turns out to have an immunelike system capable of regulating us out of existence, and does so, without us ever having truly established communication with it, while we twitter on about the man-eaters in the stars.

PART III
GAIA SINGS THE BLUES

CHAPTER 8

THERMOSEMIOSIS

Boltzmann's Sleight, Trim's Hat, and the Confusion concerning Entropy

THERMODYNAMICS STARTED OFF bright enough, practical and blond, saving the world from its limits. But then, overcome by shadows, its shiny children got dirt in their fingernails, soot in their hair; the world darkened with a foreboding of smokestacks. To the injury of overpopulation was added the attractiveness of thermodynamics as an incentive for geek speak, theoretical discussions that, with poetic justice, generated more heat than light.

Unlike economics, a different kind of dismal science, thermodynamics was an indisputable success, its application helping ignite the Industrial Revolution and its theory, in the form of Maxwell's demon, helping kindle computers and the information age. Indeed, thermodynamics may be responsible for your existence, as well as most of the nitrogen atoms in your body. In early 1912 the German chemists Fritz Haber and Carl Bosch produced inexpensive ammonia using nitrogen from the air and hydrogen gas. This in turn enabled heavily populated countries to make cheap ammonia-based fertilizers, staving off starvation on a global scale. An interesting feedback loop: technology is man-made and man now is factory-made.

Even if you are a vegan eating organic food, some 50 percent of the nitrogen atoms inside your body, including in the amino acids that make up your proteins, and in your DNA, are synthetic: they were made under high pressures and temperatures in giant factories that use 2 percent of Earth's energy, breaking the covalent bonds of nitrogen atoms in the atmosphere into forms that can be taken in by crops, eaten by food animals and us. According to Thomas Hager, these "giant factories, usually located in remote areas, that drink rivers of water, inhale oceans of air . . . burn about 2 percent of all the earth's energy. If all the machines these men invented were shut down today, more than two billion people would starve to death."[1]

Yet despite its importance, the essence of thermodynamics remains confusing. Perhaps the enormous success of thermodynamics, in both academic theory and industrial production, caused experts to ignore simple descriptions of what the second law means over the past century. The astronomer Arthur Eddington said,

> The law that entropy always increases—the second law of thermodynamics—holds, I think, the supreme position among the laws of Nature. If someone points out to you that your pet theory of the universe is in disagreement with Maxwell's equations—then so much the worse for Maxwell's equations. If it is found to be contradicted by observation—well, these experimentalists do bungle things sometimes. But if your theory is found to be against the second law of thermodynamics I can give you no hope; there is nothing for it but to collapse in deepest humiliation.[2]

Appointed to give the Sir Robert Rede Lecture on May 7, 1959, Charles Percy Snow chose to critique higher education. Snow—a baron of the city of Leicester, England, as well as a physicist, mystery writer, defender of the realist novel, and author of the seven-volume *Strangers and Brothers* (made into a BBC series)—prodded his august audience in words that were, for all intents and purposes, the first shots in what would become the culture wars:

> A good many times I have been present at gatherings of people who, by the standards of the traditional culture, are thought highly educated and who have with considerable gusto been expressing their incredulity at the illiteracy of scientists. Once or twice I have been provoked and have asked the company how many of them could describe the Second Law of Thermodynamics. The response was cold: it was also negative. Yet I was asking

> something which is about the scientific equivalent of: "Have you read a work of Shakespeare's?"
>
> I now believe [he added later to published versions of his remarks] that if I had asked an even simpler question—such as, What do you mean by mass, or acceleration, which is the scientific equivalent of saying, "Can you read?"—not more than one in ten of the highly educated would have felt that I was speaking the same language. So the great edifice of modern physics goes up, and the majority of the cleverest people in the western world have about as much insight into it as their Neolithic ancestors would have had.[3]

According to the chemist Frank L. Lambert, who has written extensively about simple thermodynamics,[4] even Snow neglected to define the essence of thermodynamics he said was so important. It has nothing to do with mandating an inevitable increase in disorder, as all sorts of cultural theorists and science geeks believe. Rather, the elegant essence of the phenomenon the second law describes is that energy, if not hindered, spreads.

The intellectual sin of focusing on disorder is wildly democratic in its choice of victims. It afflicts not only the Cal Tech astrophysicist Sean Carroll, in *From Eternity to Here,* but Pope Pius XII, who offers the second law as proof of the existence of God, because only he had the wherewithal, in creating organized life and man, to resist this all-inclusive law of ever-increasing disorder. Yet hardly anyone is safe from this widespread mistaken meme, not even the most scientific-seeming rationalists and atheists. For example, the Darwinist philosopher Daniel Dennett repeats a version of this same mistake when he writes that life-forms "are things that defy" and constitute a "systematic reversal" of the second law.[5] Superficially, this may seem to be the case, as some of life's key chemicals concentrate rather than spread energy. But it is crucial to realize that, overall, living systems spread energy and that their partial molecular concentration of energy, and production of gradients, abets this process. Saying that life defies the second law is like saying that Robin Hood is against the spread of wealth because he gives it to the poor. A watch must have its watchmaker. A car does not put itself together from parts. Nonetheless, moving atoms do join to form compounds and more complex molecules. Unhindered, energy spontaneously spreads out.

While a red Ferrari doesn't assemble itself from spare parts in a junkyard during a windstorm, this has little to do with the organization we see in life. The macro-objects of our everyday life do not behave in the same way as atoms and molecules. Car parts in a junkyard don't routinely whiz by at two hundred to two thousand miles an hour, colliding with one another, fusing and releasing so much energy that they become white-hot. Such behavior, however, is normal for molecules.

The vast majority of compounds, some quite complex, form easily. But molecules are not atoms mixed up at random like the batter around a chicken leg in a bag of Shake 'n Bake. When three or more atoms aggregate to make a molecule, they possess a precise order. Their atoms, in a relatively fixed geometric relationship, generally stay stable. When atoms "bond" after their violent collisions, they aggregate into molecules so stable that temperatures of thousands of degrees are needed to pry them apart. Melt them together (to make them move more rapidly), and amino acids form huge new compounds. The melted amino acids make "proteinoids" with hundreds to thousands of amino acid units firmly joined in the same kind of bonds that hold proteins together. The result is not useful or valuable proteins, but it does show how easily gigantic complex molecules can form naturally.

There are millions of compounds that have less energy in them than the elements of which they are composed; they are the result of "downhill" reactions, formed easily, resulting in the spread of energy. Their formation is no more mysterious than a glob of toothpaste appearing at the end of a squeezed tube.

The rules of energy science favor formation of complex, geometrically ordered molecules. But there are also compounds that require, like objects in a factory, additions of energy from outside, leading to compounds having more energy in them than they had before. Such molecules may result, for example, by molecules being energized by lightning.

While less likely, such reactions to create higher-energy molecules happen all the time. Alkanes, for example, are among the simplest of organic compounds, composed only of carbon and hydrogen, and containing portions or sections with one carbon atom holding two or three hydrogen atoms. Both simple and complex alkanes have been detected by spectroscopic methods in space. Simple alkanes with two to five carbon atoms joined to one another (and hydrogens attached to each carbon) all contain less energy than their elements. More complex alkanes, with six or more carbon atoms joined to make their molecules, have more energy in them than the elements from which they come.

These alkanes, like life's key energy-storage molecule, ATP—adenosine triphosphate, structurally a cousin to DNA—require energy to be formed. ATP is an amazing molecule. Not only is it as omnipresent in life as DNA or RNA, but because it is built up from energy and spent in metabolism, it is like cash in a casino: You synthesize and break up roughly your entire stock (about 8.8 ounces) of ATP each day.

Alkanes and similar so-called *endergonic* chemicals contain more energy in them than the elements that go into them because they are forged via input of external energy—ultraviolet or X-rays, both plentiful in many parts of the universe. This energy is not so hard to come by. Indeed, high-energy cosmic rays are penetrating your body this very moment. On average, about five cosmic rays penetrate every square inch of your body every second, right now, even as you are reading this sentence: gamma rays, X-rays, subatomic particles. Some come from the sun, but a lot come from supernova explosions.

Such abundant energy—and UV radiation was more plentiful on the early Earth—bombarded simple chemicals all the time, sometimes creating more energetic compounds that stored energy which, released later, created a cascade of delayed reactions. So even with science's splendid emphasis on connecting humans and life to natural cosmic processes, it is easy to see where the notion that life defies the second law comes from, but it is wrong. It is far better to think of living systems as temporarily deferring second law–"based" predictions of immediate gradient breakdown, but as part of a process of greater overall and longer-lasting energy delocalization. Temperature measurements by low-flying airplanes over the H. J. Andrews Experimental Forest in Oregon corroborated thermal satellite pictures showing that rain forests in summer are (because of cloud cover) as cool as Siberia in the winter. Quarries and clear-cut forests have higher temperatures than a twenty-five-year-old Douglas fir plantation and a natural forest twenty-five years old, and neither of these was as cool as an old-growth forest four hundred years old. At first one might be tempted to explain these data by saying that the capture of solar energy in the cooler versus more fallow areas is due to the buildup in them of energy-storing chemicals that prevent energy's spread. Yet there is another interpretation, which is quite different. Consider a refrigerator, keeping itself cool internally but generating excess heat. Is not this the essence of the grasslands compared with the desert, the forests compared with grasslands, and the great jungles compared with temperate forests? Most of the solar energy in the plants goes not into "blocking" the second law to make energy-storing compounds but into the thermodynamically open process of evapotranspiration. Latent heat is released as rain. Given the solar gradient, the difference between the hot sun and cold space, the cooling provided by evapotranspiration-produced clouds must, like a refrigerator, lead to energy spread, entropy production, farther out. Like natural nonliving complex systems, and our cooling machines that require an outside source of energy and dump heat into their surroundings, organisms have impressive internal structure. Yet, seen as energetic processes rather than firmly bound things and compared with less-organized regions of matter, they produce more heat, even as they keep themselves cool. They spread more energy. Constitutively open systems, they do not defy the second law. Rather, their order is connected to a more effective, elegant, and continuous production of entropy, dispersed energy.

Entropy is a confusing word. In 1854 the German physicist Rudolf Clausius combined the word *energy* with "tropos," Greek for transformation, to come up with entropy for a change in energy, dq: $\Delta S = dq/T$.[6] This was later given a statistical formation by Ludwig Boltzmann (1844–1906), one of the founders of modern thermodynamics. Boltzmann in a single statement is probably responsible for the lion's share of our confusion about the conceptual meaning of entropy. After more than four hundred pages of heavy math, in a common-language summation, Boltzmann writes (in *Lectures on Gas Theory* [*Vorlesungenuber Gastheorie*]) that the universe, "or at least most of the parts of it surrounding us are initially in a very ordered—therefore very improbable—state. When this is the case, then whenever two or more small parts of it come into interaction with each other, the system formed by these parts is also initially in an ordered state and when left to itself it rapidly proceeds to the disordered most probable state."[7]

This concept—of entropy as "disorder" and thus any type of disorder as "entropy"—was dominant throughout the twentieth century. In the equation for entropy the symbol for entropy is *S*. Boltzmann had developed an equation for the entropy change, ΔS, in terms of energy states, but he could not do actual calculations because he did not know how to discover the value of k, now known as Boltzmann's constant. Before Boltzmann committed suicide, but without his hearing about it, the physicist Max Planck established that k was equal to R/N—the gas constant divided by the number of molecules in a mole. The equation $S = k \log W$ (engraved on Boltzmann's 1906 tombstone) is actually a version of this equation, coined about 1900 by Planck.

In retrospect, Boltzmann's common-language summary after four-hundred-odd pages full of mathematics clouded the issue far more than illuminating it. Adding to the confusion was the code-breaking physicist John von Neumann, who advised Claude Shannon, innovator of information theory, which was to become the basis of global telecommunications, to adopt the term *entropy* to describe information. "Nobody knows what entropy really is" anyway, counseled the troublemaker von Neumann, a heavy drinker who had so many car wrecks that they named an intersection in Princeton Von Neumann Corner.[8] Shannon took the advice.

The notion that the key concept of entropy is a spontaneous change from order to disorder stems from this 1898 summary Boltzmann gave of his own work. But what is order? It certainly appears that a sparkling crystal of ice is obviously more "ordered" than an equal volume of water, but the difference in the numbers of energy states is totally beyond our comprehension. As Lambert writes, "If liquid water at 273 K, with its $10^{1,991,000,000,000,000,000,000,000}$ accessible microstates [quantized molecular arrangements] is considered 'disorderly,' how can ice at 273 K that has $10^{1,299,000,000,000,000,000,000,000}$ accessible microstates be considered 'orderly'?"[9]

It may take dictionaries twenty years to change, but already disorder is dead as shorthand for entropy in most first-year collegiate chemistry texts. It is being replaced with the idea that energy dispersal is the essence of the change described by the second law. Energy, if unhindered, spreads out in space and on a greater number of energy levels. This tendency of energy to disperse is the essence of the second law that Snow neglected to mention. A loud sound spreads out through the air from a speaker, a crashing car spews metal and heat in all directions, potential becomes kinetic energy if we fall from a tree. This dispersion tendency does not have to happen right away, however. It can be dammed or blocked. Indeed such complex damming and blocking is crucial to living systems, whose energetic molecules are safeguarded from immediate dispersion by their structure. Like a boy being pushed before he falls out of a tree, life's molecules require higher energy levels, so-called activation energies, before they disperse their stored energy. The metastable molecules of organisms are protected from energy dispersion, as are their evolved networks of repair mechanisms. They do not defy the second law but block its immediate action. The reason we don't burst into flames is because of Ea, the energy of activation, usually stated in units of kilojoules per mole.

We often lament our lot, crying why me? Our cultural stories about why things go wrong in our lives include Satan, karma, and Murphy's law, which states that everything that can go wrong will go wrong, with many comic variations, such as Roberts's axiom, "Only errors exist." But our acute awareness of problems, pains, and errors is itself part of living systems' protective feedback systems. Chemical kinetics in physiology continuously safeguards life from spontaneous combustion and other forms of destruction that would occur if the second law mandate of dispersion were immediately fulfilled. On the other hand, life's organized systems, chemically and cybernetically (because of cyclical, sensitive feedback loops) protect it from immediate breakdown and allow it to prolong its "entropy production." In animals, that involves the recognition of and oxygen-aided breakdown of energetically concentrated chemical substrates—food—which we continue to take in to maintain our body and mind acting in the world around us. Bacteria have greater diversity in the concentrated energy sources and chemical substrates that they can use to run their metabolism. Fungi produce enzymes that digest food outside their bodies before they ingest it. Plants, generally considered inferior to animals, in fact are metabolically superior. We learn in grade school that plants produce oxygen that we breathe, and breathe carbon dioxide that we exhale, suggesting an essential equivalence and a nice ecological match between plants and animals. But plants not only photosynthesize, producing oxygen; they also use oxygen. They do it at night when sunlight is not available as a source of energy. They do so using mitochondria, former respiring bacteria, the same inclusions we have in our cells.

All these life-forms, however ordered and protected they are from immediate gradient breakdown, are actively and profoundly engaged in energy dispersal. We do not burn up like a sparked piece of paper, but while alive we seek concentrated sources of energy such as food, methane, and oil. We pay attention when things go wrong, as in Murphy's law, but we miss the fact that at the unconscious biochemical, cellular, and physiological levels things continuously go right, protecting us as natural machines to spread energy in accord with the second law. Each of us is the result of 3.8 billion years of evolution of highly organized, actively metabolizing, energy-dispersing systems that are protected by chemical kinetics and molecular barriers. We store energy the better to disperse it. Our damming and delaying of the effects of the second law locally allow us, as open thermodynamic systems, to maintain and increase our personal realm of active gradient reduction. Rather than bemoan our lot, we should be continuously amazed at the exquisite artistry of life that has used chemical kinetics to keep us going and growing with nanotechnological precision since shortly after Earth's origin. Murphy was wrong. And although Neil Young said it's better to burn out than fade away, better still is what life has been doing, growing its domain of controlled burning to protect itself even as it finds new gradients, new concentrated sources of energy to its organized, ordered, energy-dispersing selves.

More confusion stems from entropy's status as a ratio in its original definition. A "low entropy (ordered)" state, in a typical expository article by a highly competent physicist in a *New Scientist* article,[10] illustrates the situation. The *New Scientist* article befuddlingly labels both an illustration of the big bang—indicating that in its first instant our cosmos was a seething mass containing the entire energy of the present universe but in a relatively small volume with an extremely high temperature—*and also,* three pages later, a panther who stalks a cage as "low entropy." An arrow, marked "Ordered energy and matter," enters the cage. But clearly the zoo environment and panther's food are not billion-degree seething masses of quarks and gluons. The phrase does not add to clarity, especially when the "low entropy (ordered)" state also describes a smallish box of indolent molecules at a very low temperature.

Clearly, a great advantage of introducing "entropy increase as due to molecular energy spreading out in space," if it is not constrained, begins with the ready parallels to spontaneous behavior of kinds of energy that are well-known to beginners: "the light from a light bulb, the sound from a stereo, the waves from a rock dropped in a swimming pool, the air from a punctured tire."[11]

The physicist Harvey Leff, coeditor of definitive anthologies on Maxwell's demon, writes in a pedagogical note in *The Physics Teacher*, "It is a remarkable, fortuitous coincidence that entropy's traditional symbol S can be viewed as shorthand for 'Spreading function.' Using the interpretation of spreading over space and time, entropy might become more meaningful to you and your students."[12]

In the Woody Allen film *Whatever Works*, an attractive young woman from Mississippi persuades a retired physicist (and chess teacher) from Greenwich Village to let her live with him. Her mother finds her and lets her know that she is unhappy with the daughter's choice of mate, encouraging a handsome British actor to pursue her instead. After initially resisting the young man's advances, the young woman finds herself on his houseboat, happily pushing him away after a passionate kiss. Overcome by a moment of reflection—she is still under the intellectual influence of her grumpy string theorist benefactor (who is now her husband)—she returns from her mental ruminations to gaze into the actor's eyes. What is it? he asks.

"Entropy," she murmurs, explaining that what happened was like a "tube of toothpaste," by which she means that life, like toothpaste squeezed from its tube, can never return to the way it was.

Molecules of perfume speeding at hundreds of miles per hour but constantly colliding with equally rapidly traveling air molecules will move from one side of an absolutely quiet-air room to the other. Cream molecules, totally unstirred, independently and inevitably, similarly collide and slide from the top of a cold coffee cup throughout that cup. They will always move, disperse, and thereby delocalize and disperse their energy by spreading out in simple three-dimensional space if they are not constrained. *S*—the spreading function: This simple formulation beautifully generalizes much of the ordinary phenomena in our lives. If you place a hot piece of iron on another, cooler piece of iron, "heat energy" flows to the cooler iron until the two become exactly the same temperature. Technically, the "heat energy that flows" is actually the vibrational energy of atoms dancing in place in the metal that, on average, are moving faster in the warmer iron bar than in the cooler. At the surfaces of contact, the vibrations of the warmer bar interact with slightly slower vibrations in the cooler bar, and over time the surplus energy of the warmer bar disperses. It spreads out, so the vibrations of atoms held in place are at the same energy levels in both bars.

PROTOSEMIOSIS

Simple as it is compared with the more sexy, obscure term *entropy,* the *S* function has major implications for philosophy. If we didn't know better, we might be tempted to risk the hypothesis that the bar was heated on one end and wanted its warmth to spread to its cool parts. That might not be its goal in any conscious human sense, but that is clearly its direction, its unconscious orientation.

Spreading comes before meaning, before discrete sense. We are reminded of Georges Bataille's writing that his ink is like blood, the slow spill of a lifelong intellectual suicide or sacrifice. Energy's delocalization sets the stage for teleology, because it has a natural end, equilibrium, and will find ways, sometimes complex ways, of getting there.

The brute reality of this protoconscious, protosemiotic process ultimately poses Copernican inferiority problems for a certain not-so-hairy prodigy and problem-child great ape species; let's not mention any names. The main problem seems to be that we like to congratulate ourselves for being the sole full possessors of a certain sort of truly teleological purposiveness. Some see glimmers of sign making and purposive behavior in other creatures, and biosemioticians may grant the power to all life, even considering it its distinctive feature. But in general we conflate purposiveness with human consciousness—not noticing that it is implicit in the telic substrate of a thermodynamic universe "prior to" or "irrespective of" life. The phenomenologically observed tendency of things to go from being concentrated to spread out demarcates a natural telos, and the relative end-state of "being spread out" (thermal or chemical equilibrium) seems to select for, tug, or pull random aggregates to become more organized to accomplish that natural end.

It is no coincidence that plants, animals, fungi, and bacteria spread as they grow or that, even if they are not growing or reproducing but merely metabolizing, they put more entropy, mostly as heat, out into the local environment than would be the case without them. The organism itself, from όργανον—*organon* in Greek, meaning "instrument" or "tool"—seems to be a kind of natural device, a kind of cosmic organ for the degradation of the ambient concentrated energy sources that it craves. At least unconsciously, organisms must recognize the ambient gradients that support them and that they degrade.

BOLTZMANN'S SLEIGHT

In Isabelle Stengers's judgment, "Boltzmann was forced to recognize that his theorem did not describe the *impossibility* of an evolution that would lead to a spontaneous decrease in entropy and would, therefore, contravene the second law of thermodynamics, but only its *improbability*."[13] Spreading occurs as radiation, as the big bang, as change that does not return in linear time. But this is confounding to the mathematical mind, which imagines eternities, symmetries, and geometrical perfection. Boltzmann's trick of deriving experiential, time-flowing *thermo*-dynamic spread from time-reversible, symmetrical *dynamics* is a kind of mathematical sleight. But just because there are more ways in which things can be disordered than ordered does not mean that they have to go in that direction, as over infinite time even the most orderly combinations will recur an infinite number of times. So, too, assuming that the past is ordered and the future as disordered, as Boltzmann did, when the difference between past and future is precisely what we are trying to explain, is problematic. The problem is so difficult that apparently Albert Einstein and Kurt Gödel gave up on improving upon Boltzmann's derivation of linear time from time-independent mechanics after which the former devoted himself to more tractable pursuits, developing the special and general theories of relativity.

Many times in the history of science there has been a fatal disconnect between the insular certitudes of the mathematical realm explored by theorists. The unifier of electricity and magnetism, James Clerk Maxwell, had a name for such idealism: he called it "the Queen of Heaven," meaning roughly that what his German mathematical colleagues wanted they would never have. The term is from a letter in December 1873 from Maxwell to his friend Peter Guthrie Tait, in which Maxwell derides the attempts of Clausius and Boltzmann to reduce irreversible nature (as described by entropy, the second law of thermodynamics) to the reversible equations of mathematics: "The *Hamiltonsche Princip* [i.e., the Hamiltonian Principle, a mathematical formalism that is applied also today in quantum mechanics], the while soars along in a region unvexed by statistical considerations while the German Icari [i.e., Clausius and Boltzmann] flap their waxen wings in *Nephelococcygia* [Νεφελοκοκκυγία, Cuckoo Cloud Land—referring to Aristophanes's play *The Birds* in which the trusting and naive characters, done with Earth and Olympus, plan to build a perfect city in the clouds] amid those cloudy forms which the ignorance and finitude of human science have invested with the incommunicable attributes of the invisible Queen of heaven."[14]

For Maxwell these German physicists were seeing the mathematical equivalent of animals in the clouds—they saw them all right, but they weren't really there. Mathematics, for example in Newton's theory of gravity, which ultimately allowed the Earth to be reflected as faces in the visors of cosmonauts, can sometimes be spectacularly successful. But despite the power of math on its own, its application to physics is not a lock. Richard Feynman quipped that mathematics is to physics as masturbation is to lovemaking. (He also said physics is like sex: it gives results but that's not why we do it.)

"You are a magician," an admirer tells the ballet impresario in the 1948 film *The Red Shoes*, complimenting him for putting together a great show based on the Hans Christian Andersen fable on a shoestring budget in so little time.

"Ah, yes," says Boris Lermontov, "but even for a magician to pull a rabbit from a hat, the *rab*bit must already be in the *hat.*" In the case of the road from time-reversible dynamics to the lived world of thermodynamic spreading, there doesn't seem to be a way to get from there to here, where we always end up except in our mathematical abstractions, geometric projections, and imagined infinitudes.

ISABELLE STENGERS

Isabelle Stengers crosses the two cultures, scientific and human, the mechanical-reversible and the irreversible-real. A student and colleague of Ilya Prigogine, who won the Nobel Prize in physics for inroads in the problem Einstein abandoned, she and Prigogine wrote *Order out of Chaos,* a book that attempted to introduce temporal irreversibility into the heart of physics and that developed the very important notion of dissipative structures, real "live" (i.e., not computer programs) three-dimensional systems that appeared in energy flows. If they continued their access to energy and matter substrate, they could continue to grow, producing entropy and spreading energy. In Prigogine's descriptions they were "far from equilibrium," and the farther they evolved or diverged from an equilibrium state, the more sensitive to external conditions they became. Eventually, they could undergo bifurcations, separating to new, perhaps higher-energy meta-stable energy regimes.

I met Stengers briefly in passing twenty years ago, at the annual festival of Spoleto. I didn't know anything about her except that she was a scientist. Spoleto, with its small, crooked cobbled streets and outlying fields of giant sunflowers showered by rays of Renaissance master light, is a beautiful, picturesque villa. My son was a toddler at the time, and a tricycle-possessing pixie with big blue glasses, the grandson of a fascist, was showing my son how to ride. My mother introduced me to Stengers as summer air wafted over the stone balcony. Not long after, Umberto Eco could be seen in the corner, smoking a cigar and surrounded by Italian reporters as if he were Marcello Mastroianni beleaguered by paparazzi in a Fellini movie. Eco had just published *Foucault's Pendulum,* which was everywhere in the airports. And although we had just met her, Stengers suggested my son, Tonio, might want to spill some water on Eco.

As luck or Gaia would have it, a few years after meeting her I began working, at the behest of the Montana ecologist Eric Schneider, on popularizing the thermodynamics of complex systems—especially as applied to life, which was not Prigogine and Stengers's focus. Two decades later I found myself at a small conference celebrating her philosophical work *Cosmopolitics: In Place of Both Absolutism and Tolerance.* I was invited by Eben Kirksey, it was put on by the Mellon Committee for Science Studies at the Graduate Center of CUNY, and my offering was "Time Tricks, or Lure of the Retrocausal: Isabelle Stengers' Delicate Operations on the Body of Science."

A week before the conference, I was reminded of the strange interweaving of the threads of our lives in the folding tapestry of time by an e-mail, which I forwarded to Stengers, reminding her also of our original meeting when she advocated that my son douse Eco. I also forwarded it to other experts in thermodynamics.

The e-mail was from a Dr. Graf in Germany.

"This April 17th," it announced, in what could pass for a digital-age version of a broadside for a traveling medicine show, "Dr. Graf will be demonstrating his suitcase-sized gravity machine, a perpetual motion machine of the second kind." Curious, because I knew the U.S. Patent Office no longer even accepted applications for perpetual motion machines, I fired an e-mail off to Frank L. Lambert. Lambert e-mailed me back to say that Dr. Graf was no mountebank. He had looked at Graf's calculations, and his valise-sized gravity machine looked like it worked. It did collect energy from gravity. But it was impractical. Lambert calculated that it would require a tube of 90 trillion miles to produce enough energy to light one 100-watt lightbulb.

AROUND 1990 I started working with Schneider, who had already been studying the intersection between nonequilibrium thermodynamics and life for twenty years. This led in 2004 to the University of Chicago Press book *Into the Cool: Thermodynamics, Energy Flow, and Life.* It is a post-Prigoginian work. And it focuses on life, how characteristic patterns appear in and as organisms and ecosystems, on the specifics of life itself—which was not Prigogine's focus, although some, such as the Austrian astrophysicist Erich Jantsch, who dedicated his book *The Self-Organizing Universe* to Prigogine, saw immediately a relationship between Prigogine's work and James Lovelock's description of Earth's surface as physiological.

The most basic point is that life and the second law, entropy and complex systems, are in no way opposed to each other. Far from it. Natural complex systems cycle matter in regions of energy flow. These systems are not in thermodynamic equilibrium, but, living off of and leveling ambient gradients, they help foment equilibrium in their surroundings. Telic, energy-seeking, sensing life-forms belong to a larger class of dissipative structures. Rather than become less organized, these systems become more organized as energy flows through them. Schneider provides an alternate to Lambert's modern definition of the second law. Schneider's version is that nature abhors a gradient, a gradient being a measurable difference of temperature, pressure, chemical concentration, and so forth. Living systems are among the class of natural systems that actively reduce gradients as they maintain their internal organization, grow, and spread.

And here from another angle we can espy the deep roots of teleology, the telic protosemiotic that rules nature. Nature "doesn't like" differences or concentrations of energy and will actively find ways to get rid of these differences or delocalize these concentrations. Convection cells called Bénard cells are hexagonal fluid flows. They are not less but more organized than simple conducting fluids and perform their natural task, reach their natural telos of energy dissipation, more effectively. Complex systems from autocatalytic chemical reactions similar to the precursors of life to cycling storm systems that rectify atmospheric pressure gradients have a naturalistic function, to delocalize energy. When they are finished with it, they are done, both with the tasks and with themselves—they disappear. Complex systems from Bénard cells to living organisms and ecosystems measurably increase rates and regimes of energy dissipation. Stengers confirmed this for me at the conference. Yes, it's true, she said, complex systems produce more entropy, but your story is not my story.

I showed her my Maxwell demon trick. Three red cards, representing heat or fast-moving atoms, are mixed with three black cards, representing cold or slow-moving atoms. Focusing on the little angels on the back of Bicycle brand playing cards, I turned the symmetrical cards end for end. I then showed that, with no work being done, the cards had sorted themselves out back into a red and black state. The trick simulates moving from a "more probable" state to "less probable" state, one in which a gradient, here represented by red and black cards, is reestablished. In fact this is just a modern version of a classic card trick called "Oil and Water," where red and black cards (oil and water), after being mixed, resort themselves.

It then turned out that Stengers, a fan of magic, was reading *Sleights of Mind,* a book on the application of magic tricks to cognitive psychology written by three psychologists. My belief is that, after my talk, when she said "I would rather listen to you than physicists" but "my story is not your story," it was because her passionate project is to review the history of science, to see what it could have been and would have been and what, correlatively, it could still become. A kind of pixie, she, like Jacques Derrida, productively bothers many, if not all, received notions.

Nonetheless, she and Prigogine did not work directly on living systems, and to me one of the most exciting applications of their work is to organisms, which both unconsciously (physiologically) and consciously spend their days recognizing and degrading gradients. "It is true that they produce more entropy." Stengers's confirmation bears repeating: The "complex systems" such as your child and your mother and your dog and your cell phone—which seem *more* ordered, *more* organized, and which dazzle us with their elaborateness and agency, their intricacy and relative autonomy—are actively spreading energy into the environment. They are converting high-quality energy to low-quality energy at a pace faster than would be the case if they did not exist. Nonliving systems such as Bénard cells also measurably produce more entropy. They seem to occur because nature happens upon ways to efficiently get rid of preexisting gradients. Life is a splendid means of doing so and often has fun (e.g., eating) in the process. It is something nature likes to do, with or without life and with or without consciousness—though both life and consciousness seem to increase the efficiency, partaking of the "forness" in this natural tendency—to find and efficiently delocalize preexisting concentrated energy reserves. It does thus seem that, beyond any *particular* advantage of a bodily organ, life as a whole exists *for* something—to spread energy, to delocalize and deconcentrate it, a valuable ability connected to its metabolism and franchised by its genetic reproductive capacity.

I thus disagree with the biosemioticians (e.g., Thomas Sebeok and Jesper Hoffmeyer) who would say, quite liberally given the general climate, that the realm of life coincides with that of semiosis, meaning making.[15] I think this is true and love Butler's discussion of the dog looking at a door to go outside not as language (from French *langue*, "tongue") but as "eyeage." Yes, other organisms speak, not necessarily with their tongues, and we can sometimes "hear" them, not necessarily with our ears. But language is a material mediation. It takes time to convey a message; it requires intermediary parties. And behind discrete messages I would argue is an ur-message, a message that in our heart of hearts we both already know and know that we don't want to know. At the limit that message is this: that we are among the tools nature is using to send it. Like the heat-curling tape in the old *Mission Impossible* TV series, which destroys itself after telling the spies their current mission, our ur-message is that we are to foment equilibrium and that we should destroy the message that tells us this message, getting lost in time and stories along the way. Here then, from a 1960s TV show, is a metaphor for the thermosemiotic ur-message. The secret desideratum of all living things, their "higher" message, which is also their "lower" message, and is itself beyond discrete meaning, is to reach equilibrium. This is our impossible mission.

TRIM'S HAT

"I live on Earth at present," writes Bucky Fuller, "and I don't know what I am. I know that I am not a category. I am not a thing—a noun. I seem to be a verb, an evolutionary process—an integral function of the universe."[16]

In my view that is exactly correct: We have a natural function, fomenting equilibrium, although it is complicated by the fact that completion of this prosaic task would render the message complete and the message sender obsolete. Life's solution thus is, as Schneider says, to relight the candle: reproduction continues the metabolic process of gradient reduction. Genetic replication (dependent on breaking the triply covalent bonds of nitrogen locked up as N_2 in the atmosphere and integrating them into DNA) is part of an expanding 3.8-billion-year-old process that locates and uses—without necessarily using up—available energy. It is a natural process, not defying or violating anything. And complex, sensing systems seem to have a leg up in the ancient gradient-reduction game.

The natural unconscious tendency to spread energy, enabled and enhanced by a natural creation of nonliving—and ultimately living and uncontroversially "teleological" conscious complex systems—confers a broad purposiveness over nature. It is a bit of a disappointment, however, for those looking for an ultimate metaphysical or traditional Judeo-Christian purpose, for example, to do God's will.

Tao 道, "the way" in Chinese philosophy, advocates a less humanocentric and more thermodynamically resonant view, I think, a kind of principle of least action, a kind of advocacy of aligning oneself with natural flows, of which one is part but which take part without one in any event—a philosophy that reminds me of setting goals but then opening oneself up to a path of least resistance, a difficult–easy path, free of excess interpretation, which in turn reminds me of Charles Bukowski's tombstone, which says, "Don't Try."

While life's desires are ultimately focused more on devouring chemical and photosynthetic gradients than on using gravitational ones, the tendency for things to reach their end in the natural world is something we understand, intuitively.

In 1759 Laurence Sterne, in *The Life and Opinions of Tristam Shandy, Gentleman,* has a character, Corporal Trim, who drops his hat:

> "Are we not here now"; continued the corporal, "and are we not"—dropping his hat plump upon the ground,—and pausing, before he pronounced the word—"gone! in a moment?"—The descent of the hat was as if a heavy lump of clay had been kneaded into the crown of it.—Nothing could have expressed the sentiment of mortality, of which it was the type and forerunner, like it;—his hand seemed to vanish from under it;—it fell dead;—the corporal's eye fix'd upon it, as upon a corpse;—and *Susannah* burst into a flood of tears . . . ,—meditate, meditate, I beseech you, upon *Trim's* hat.

The world may be meaningless, yet have a direction. It may have a story, but not the story we want to hear.

When Sterne tells us of the hat that "his hand seemed to vanish from under it;—it fell dead," we are not sure what he means; when was his hand ever under the hat, did he mean the hand fell dead or the hat, and if the hand, did it stand for the head? Or perhaps he meant to convey Susannah's view, and from hers the drop of the hat obscured the corporal's hand, which was behind him after the fall of the hat, creating the disappearance effect. We mistake agency, jump at our shadows, find patterns, and attribute motives after the fact even to ourselves. Language itself is a teleological nexus that, with its telic prepositions, its "fors" and "tos," draws us in; it pulls us far more than we push it. English, for example, is subtle, supple, a deep and accommodating crowd-sourced reserve of wisdom that we do well to follow, along ancient paths of polysemy, rather than try to impose on it with academic jargon and neologisms. The telic character of language brings forth its own stories.

A vivid example occurred in the notes I made for a biosemiotics conference at the Rockefeller Medical Center on Front Street in New York City. I wanted to talk about how biosemiotics, with its emphasis on language, forms a welcome bridge between the humanities and the science, but that talks in the former are generally read whereas scientific talks are usually spoken. I wanted to say that I found this difference ironic or paradoxical because it should be the other way around: scientists who assume language is a transparent medium to convey unitary knowledge would want to be more careful by writing what they intended to say, while humanities types aware of the inconstancy of language and its irreducible metaphoricity should not worry so much. Practically speaking though, I would like to be able to do both: reading a text can put people to sleep. Thus in my notes, wanting to make the above point extemporaneously, I wrote:

Reading Humanities vs. *fall*
as leep
safety net

The break between *as* and *leep* was not a typo nor was *leep* a misspelling. Rather, my handwriting had a discontinuity in the word *asleep.* Still, reading my own writing, it seemed to say something else entirely, a kind of haiku in which someone falls asleep, but the fall that takes them there turns into a leap (misspelled). In fact, my note was merely a prosaic reminder to tell people that although I would prefer to ad lib (as is more common in science talks versus humanities presentations), a written text was for me a safety net against losing my way in the forest of extemporaneity. I didn't write out the whole note so as to force myself to speak it. But when I went to look at my note before my talk, the "safety net" in the written text looked like the place at the bottom of a poem where the sleeping faller would be caught in a net of dreams.

"Fall as leep": what I had not tried to do looked more purposeful, and certainly more beautiful, than what I had.

BIOSEMIOTICS

Those still smarting from Snow's scolding may want to look to biosemiotics. Biosemioticians are interested in healing the Cartesian cut, the rift between body and mind that overlaps roughly with that between the humanities, concerned with thinking, words, people, spirit, literature, and art, and the sciences, concerned with matter and things.

So what is biosemiotics? According to Wikipedia, biosemiosis is "sign action in living systems." But I am arguing that semiosis, or the natural telos from which it forms and which structures it, is more prevalent than life, that semiosis exists in an unconscious, inchoate form in complex systems generally.

"Particular scientific fields like molecular biology, cognitive ethology, cognitive science, robotics, and neurobiology," Wikipedia continues, "deal with information processes at various levels and thus spontaneously contribute to knowledge about biosemiosis [which] may help to resolve some forms of Cartesian dualism that is still haunting philosophy of mind. By describing the continuity between body and mind, biosemiotics may also help us to understand how human 'mindedness' may naturalistically emerge from more primitive processes of embodied *animal* 'knowing.'"

Biosemioticians are interested in pursuing the rich tapestry of signification systems in living organisms without foregoing, as might be said of Continental philosophy, a reality beyond and independent of such systems. There is an interesting historical connection here as Heidegger, who is so influential in Continental philosophy, was deeply influenced by a figure foundational to biosemiotics, the protobiosemiotician (as Sebeok called him), Jakob von Uexküll. But there was a fork in the road of Uexküll's influence, as Heidegger, whose notion of *Dasein* seems largely lifted from Uexküll's notion of an organism's life-world, or *Umwelt*, turns it into something especially human.

Heidegger argues that man exclusively dwells in the house of language, that Friedrich Hölderlin is wrong in thinking that animals, even a lark flying across the sky, have access to "the open"[17]—that is, Being as a whole rather than an environment. Heidegger asserts that the human hand is vastly different from a monkey's paw. Heidegger considers all nonhuman animals together. Apparently almost any animal will do—he chooses a bee—to stand in for the essence of "the animal." Heidegger, who mentions Uexküll more than any other scientist, and though he gets the idea of *Umwelt* from him, says that animals are poor in world, separated from us language-dwellers by an "abyss."

So Heidegger takes Uexküll in a distinct direction, running with his important exploration of meaning making, purposeful behavior, and the inner worlds of various organisms, but then applying them almost solely to man. Biosemiotics takes a different fork in the path, expanding Uexküll's insights to all animals and even to the chemical process of life itself. Wendy Wheeler, for example, in her book on biosemiotics and culture, cites approvingly Hoffmeyer's notion that life's basic unit may be the sign and that semiosis and living coincide precisely.[18]

I think that biosemiotics does provide a way not to deny mind, or mindlike processes, in nature. I agree with this spirit of connectionism exemplified by biosemiotics in its effort to exorcise, with profound realism, the ghost of Cartesianism that haunts science. Jablonka et al. describe heredity, signification, as "four-dimensional": genetic, epigenetic, behavioral, and human-style symbolic messages are sent across the generations.[19] In epidemiology and psychoneuroimmunology, there is evidence that declines in our social status can shorten our lives, presumably through some sort of signal, that our attitude can affect our immune system, sickening or strengthening us depending on the conscious and unconscious messages we send ourselves. Candace Pert identified peptide receptors in the brain and shows how chemical messengers pass between the endocrine, nervous, gastrointestinal, and immune systems.[20]

Sign systems work not only inside our bodies in cells but outside them in ecosystems. We now know that genes exist that ensure timely aging unto death. Apoptosis, telemorase rationing, and sugar intake are all involved in ensuring death in aging organisms. (Not all organisms age, which seems strange to us because we do, but, if you think about it, life has already "figured out" the way to maintain as a complex system, basically by directing energy to genetic replication—"restarting the flame" as Schneider says.) The signals for aging are partly mediated by food intake, which is why near-starvation diets allow animals to increase their health and life. Josh Mitteldorf and John Pepper argue that aging behaves like a program that maintains genetic turnover and stabilizes population flux in ecosystems subject to crashing. This program seems to turn off when populations are low on food, which is why organisms of various species can live longer and even become disease resistant by lowering their intake. Caloric restriction is thus an example of ecosemiosis. The ecosystem talks directly to our genes, normally without us consciously realizing it, and science is just beginning to overhear this fascinating conversation.[21]

Clearly semiosis and life are connected. *But* I would argue that the basic signification we see in responsive life, with its naturalistic teleology—the plant reaching for light, the animal shivering to stay warm—has deep roots in the *non*living world. An insulator who uses powdered dust to watch air escaping in a leaky house told me of his surprise to see a streamer go halfway across a ceiling before doing an about-face and returning through the holes of the electric outlet whence it came: even near-equilibrium systems with nothing an ordinary person would remotely call alive seem to "figure out" how to accomplish their natural telos. Prigogine and Stengers, in discussing the coherence of an only slightly more complex thermodynamic system, a Bénard or convection cell, discuss how the parts seem to be "communicating."

Hoffmeyer writes: "This inversed arrow of time (future directedness) immediately sets functions apart from other kinds of mechanisms that always refer backward along some chain of causation [in] explaining how the feature occurred."[22] I believe this to be the crux of the issue, as humans conflate function, sign-making processes (semiosis and biosemiosis), and conscious or "purposive" processes.[23] Immanuel Kant, in the *Teleology of Judgment,* showed that teleology (like causality, space, and time) is a mental category that we bring to the world. Indeed, it is not so easy to distinguish *consciousness* from this kind of "future directedness [that] immediately sets functions apart from other kinds of mechanisms [that instead] refer backward along some chain of causation"—"mechanisms" here referring to phenomena such as natural selection and Newtonian-style action-reaction.[24] While it may be anathema to biosemioticians to extend sign-making behaviors beyond the realm of the living, consider the evidence: Nonliving complex systems such as hexagonal-shaped thermal convection cells, intricately changing chemical (e.g., Belousov–Zhabotinski) reactions, and typhoons multiplying over the Pacific also originate, maintain, and grow only within gradients (that they implicitly—and semiotically—sense). The differences are in temperature, chemical concentration, electron potential, and barometric pressure. Hurricane wind speeds, part of cyclically organized storms (to which humans, granting kinship, give first names), are directly correlated to atmospheric pressure gradients.

Whether we like it (in the sense of finding it flattering to our vaunted sense of human specialness) or not, such behaviors, whose natural teleology (or purpose) is to reduce ambient gradients, are genuinely *future-directed.* But, contra Kant and modern self-organization theorists, such future-directedness does not so much "emerge" as *inhere* in the fundamental telic nature of energetic matter, as described by thermodynamics' second law.[25]

In *Reading for the Plot: Design and Intention in Narrative,* Peter Brooks talks about the "Freudian master plot": apart from, behind, and informing all the complex meanings of a text or novel is a movement toward equilibrium, which for us as *individuals* means death.[26] I think we have to bracket the conflation of awareness with meaningful, patterned behaviors to understand what is going on here. *Living is a form of extending the energy degradation process,* and I don't think it takes anything away from us to see how our behavior grows out of this higher or lower realm of energy transformation and data processing, which may or may not itself have a deeper meaning. The death wish, as I explore briefly in the conclusion, stems from this protosemiotic or ur-semiotic drive to return to equilibrium. But living, which makes use of death, is even better, as it sustains and expands the open systems that produce more entropy.

CHAPTER 9

LIFE GAVE EARTH THE BLUES

NATURE IS NOT JUST RED IN TOOTH AND CLAW but green with symbiotic chloroplasts, yellow with chrysophyte algae, and flamingo-pink with ingested carotenoids. It is an amazing psychedelic display of spiraling foraminifera, radiating radiolaria, and diatomaceous earth-making diatoms. It is not just hemoglobin red with the blood of animals but nacreous and jeweled with strange partnerships, luminous microbes illuminating deep-sea animals, floating cathedrals of calcium and silicon, oceans full of miniature filigreed and fragile pillbox, star-shaped, and coin forms. On land, hordes of green beings alchemically transform sunlight and dirt and animal exhaust into fruit and flowers and, at another remove, lovers and meat, their shining, glistening, mutually orgasmic bodies a billion-year refrain of triumphant partners, a buoyant rejoinder to chromatic oversimplification, a multicolored splendor. Life is not all roses, but neither is it the opposite. A more profound poet than Lord Tennyson, William Blake, said, "Exuberance is beauty" and "Energy is Eternal Delight."

EARTH IS NO ORDINARY PLANET. We may pride ourselves on our scientific instrumentation, our thermal satellites and X-ray diffractometers, our magnetic resonance imagers and gas chromatographs determining the atomic composition of crystals and the chemical composition of stars. But the biosphere uses more ancient, distributed self-growing and self-repairing instruments to recognize and maintain its manifold operations. Global humanity is a modern variation on an ancient theme. The biosphere builds an endless variety of biomolecular concentrators and redistributors, organo-devices such as water scorpions (family: Nepidae) with built-in fathometers, plants with gravity sensors and exquisite animal behavior–modifying compounds, algae with barium sulfate and calcium ion–detection systems. Magneto-sensitive bacteria detect true magnetic north, homing pigeons and bees fly home on cloudy days. Electric fish generate and sense, via electroreception, magnetic fields, which they use to locate and communicate with one another.

It is a psychedelic planet and life lights it up. It parties with fireflies, luminous fish, glow-in-the-dark algae in Vieques Island's bioluminescent bay, and *Gonyaulax* that flash circadianly wherever they are. Green plants, red algae, and cyan-colored bacteria join us and most animals in perceiving the visible slice of the electromagnetic spectrum, which extends from 400 to 700 nanometers in wavelength, and which we see as the colors of the rainbow spanning from purple (the shortest wavelength) to red (the longest). But pollinating insects detect pretty patterns of petals visible only to those that espy within the ultraviolet range at wavelengths below 400 nanometers. Honeybees navigate by polarized light. Pit vipers such as rattlesnakes track their warm prey via infrared. Dogs detect ultrasound; bats not only detect but emit it at ultrahigh frequencies, some 100,000 cycles per second. Together we living beings make and sense and alter the composition of the soil, ocean, atmosphere, and even lithosphere, where microbes that live in the rocks, endoliths, live. Everyone alive has a history of uninterrupted life that goes back for more than three billion years. That includes you, but it also includes inchworms and *E. coli*. In that sense, there are no extant "highest" or "lower" organisms: although they may not act like it, each present-day life-form represents a successful track record of some 3.8 billion years of evolution.

The numinous feeling of aliveness we get in seeing our blue sphere from space finds support in multiple lines of evidence that show that global life is *physiological.* The continuous use of matter and energy by multifarious living beings combines to impose regularities and boundaries of action on the planet's oceans, land, and air. Biospheric life is not an organism, technically, because organisms don't completely recycle all their available atoms; they share them with other organisms in ecosystems. Earth is thus a kind of superorganismic being or, more academically, a global ecosystem. A kind of closed causal nexus, the blue planet not only reacts but responds, including to its own plentiful inhabitants/constituents.

The notion that Earth is a rock with some life on it is part of our historical heritage and an example of Cartesian dualism: here, on the planetary *rock,* we have matter; while over here, in moving organisms, we have *life.* But life and its environment are so tightly wound, it's wrong to speak this way. In fact, life and the environment form a single system, an energetic phenomenon of chemistry and movement, connected to the sun, at Earth's surface. "Life" is a kind of substance, one we feel from the inside, but which consists of cosmically available elements organized in regions of energy flow. From a materialistic standpoint, living matter is a peculiar moving mineral made mostly of water. It is, indeed, an impure form of water.

"I am as pure as the driven slush," said Tallulah Bankhead in that louche and ultimately cosmic quip. It's no insult that she boasted of her impurity, her promiscuous materiality. The processes of living organize many minerals on Earth's surface. Human teeth, for example, are converted toxic waste dumps: Evolutionarily, my teeth derive from the need of marine cells to dump calcium waste outside their cell membranes. Calcium is a mineral that will wreak havoc with normal cellular metabolism. Trucking this hazardous waste across cell lines in ancestral colonies of marine cells may well be the basis of all present-day shell- and bone-making, including the apatite, a combo of calcium phosphate, calcium carbonate, calcium fluoride (not the same as sodium fluoride added to drinking water), and other compounds, in a smile. Prevalent calcium minerals such as calcite, aragonite, carbonate, phosphate, halite, gypsum, and so forth are also the dominant media used in biomineralization. Opal, a semiprecious type of silica known for its iridescent play of colors, comes next. The magnetic mineral magnetite caps the teeth of chitons but is also found inside the cells of bacteria, in swimming forms of algae, and in minute quantities in the brains of migratory fish, birds, sea turtles, and honeybees where it may act as a compass. Opossum shrimp use needle-shaped fluorite crystals to avoid the light. Like the found objects a junk artist turns into beautiful works, calcium ions once poisonous to marine cells are now arranged in crystalline lattices to make shells and bones including those of our ancestors with spines and central nervous systems, including, that is, those of the junk artist himself. Beautifully symmetrical marine microbes known as radiolarians deplete the oceans of amorphous silica and strontium to produce their ornate skeletons. The dried leaves of a New Zealand shrub, *Hybanthus floribundus,* contain up to 10 percent of nickel, a greater percentage than some mineral sources currently being mined.[1] The concentration of vanadium in marine animals known as ascidians rivals the concentration of iron in ours. The chambered nautilus, related to the squid and octopus, has a powerful aragonite "beak" capable of crushing bones. Its shell is also aragonite, while its balancing organ is formed of calcite crystals in a "pinpoint mineralization"; and it has normal kidney stones made of phosphate minerals, inclusions shared even by nautilus-type organisms whose kidneys have lost their function. Their inclusions continue to serve, however, as reservoirs used to store the raw skeletal materials calcium and phosphate within the organism. Mediating cell-to-cell interactions as part of the putative neurological basis for motor reflexes and thinking, calcium is lethal to cells in a free ionic state; but although calcium ions are ten thousand times more prevalent in the oceans than is the poison cyanide, calcium itself has been incorporated into the very marrow of life, into its skeletons and shells, and in physiological processes ranging from blood clotting to thinking.

The elements incorporated into the psychedelic biosphere's flowing functional design include silicon, the second most common element in Earth's crust after oxygen. Ninety percent of the minerals in Earth's crust are silicates. As silicon dioxide, this element composes glass, agate, tiger's eye, and rose quartz. The technology industry has found it useful, as it is used in the silicon chips of our computers and cell phones. But if we recognize it in the stained glass of our churches, we should also note it as the tiny beautiful exteriors of diatoms (which can look like stained glass!), radiolaria, sponges, and other organisms. And silicon is only one example. Life has been biomineralizing its environment for thousands of millions of years, turning its house into a home and its home into a body as it assembles ever more complex bioarchitectures from its "nonliving" surroundings. Clarice Lispector, who said she felt happy only when she was writing—that is, being a conduit between the graphite tip of a pencil and the photosynthetically produced surface of a page or, rather, in her case, being the incarnadine agent that touches the key of a metallic typewriter that imprints and scars earthly matter in a memorable way—remarked that when she was young, before her mother died so young, books to her appeared to be natural things, like fruit or babies. In a geophysiological sense we must admit that the little Clarice was right: an extraterrestrial, looking at Earth's mineral flows without access to the awareness of the beings within those flows, would no doubt consider books another example of Earth's boggled body of transformative biochemical processes. We come out of Earth like books come out of us; technology and biology are different aspects of a single process.

THE ENTIRE COSMOS is not composed mainly of water, but is made mostly of one of water's ingredients: hydrogen. So is life. The hydrogen of the universe appeared from the beginning in the big bang some thirteen billion years ago. Hydrogen, the basic stuff of the entire visible sky, is converted to helium and other heavier chemical elements in the center of those natural nuclear reactors known as stars. We recognize hydrogen on Earth mostly in its combined form as the elemental component, the one that besides oxygen makes up water. Yet some hydrogen on Earth does persist in a purer form; it is a colorless, odorless, lightweight gas (H_2) that, as in a balloon, easily escapes from Earth's gravity. Life, when active, is always composed mostly of water. But life's ubiquitous H_2O, water, is not the elusive H_2 (hydrogen) gas. H_2, the stuff of stars, makes only a cameo appearance on Earth. Gas hydrogen, when sparked, is violently reactive; to find it in nature one must crouch in the mud or descend into the smelly depths of the Black Sea. For hydrogen's energetic reactivity is valuable, and, unlike water, it is easily converted into organic matter by particular kinds of life under the Sun's energizing rays. "Organic matter" is merely a simplified name for millions of chemical compounds, many of them foodstuffs, made of hydrogen bound to carbon.

On our arm of the galaxy, the Sun and its companions ignited from a cloud of gas concentrated by gravity over four billion years ago. At that time, most of the hydrogen atoms—those that failed to remain in the Sun—escaped to the outskirts of our solar system. Today hydrogen exists mainly as cold gas and hydrogen-rich ices (of methane, CH_4, and of ammonia, NH_3) of the outer planets (Jupiter, Saturn, Uranus, Neptune) and of their moons that retained gaseous atmospheres. But here, in the inner solar system, the bodies of the rocky planets, Mars, Earth, and Venus, have not been massive enough to retain this atomic constituent of water, this lightest of elements. Hydrogen, the light stuff of stars, mainly has escaped to outer space from Mars and Venus, our planetary neighbors. Here on Earth, hydrogen lies hidden in one massive disguise: inside the three-thousand-meter layer of water on the surface of our planet. (The average depth of the world's oceans is three kilometers.) Mars, too cold, and Venus, too hot, both lack any open bodies of liquid water and have retained, on their surfaces, less than a single meter of water as vapor or ice!

Although our three kilometers of hydrogen have had billions of years to escape, we suspect the element has remained tenaciously on Earth for only one reason: the incessant thirst of highly active living matter. Life, which originated in water, remains composed of water. Cyclically making itself and remaking itself, life may be as much the reason that water remains on Earth as water is the reason why life remains on Earth.

WHEN HUMANKIND ACCOMPLISHED ITS EPIC GOAL, landing on the moon in 1969, what was not immediately seen was that the territorial move to conquer space for America had a windfall pointed in the opposite direction: that of the planet we left, rather than of the satellite that was visited. As in a trip, sometimes the most fateful and educational part of the journey is the homecoming. It may have been impossible to see that Earth's surface formed a single ecosystem before landing on the moon or investigating to see if there was life on Mars. But in retrospect we realize that our view of our home has forever changed: we now see the artificiality and anthropomorphism of the European colonial view of Earth as a color-coded globe divided into nation-states—and the relative magnificence of a profound blue orb, engulfing us but itself a tiny drop in the immensity of space. And we exist as a tiny part of that tiny drop, with great potential perhaps but no proven staying power.

Over 99.99 percent of the species that have evolved on the Earth's surface have become extinct. The longest-living ones are those that entered other cells, providing them with services such as the ability to metabolize oxygen or the power to use light to make food that came from symbiotic transformation. For us to make the transition to a long-term viable life-form is by no means assured. But if we do, we'll probably have to be like those symbiotic transformers that formed close alliances with other, very different beings, lending them our special skills to make a more powerful union. In retrospect we realize that life does not just exist on Earth like a snake, say, slithering over the face of a rock. Rather, life exists not just on Earth but *in* the Earth and *as* the Earth's planetary surface, including its oceans and atmosphere.

It has been suggested that the continuous presence of water on the Earth's surface for the last four billion years is not an accident but the result of life, which is made mostly of water and continues to maintain itself in (and as) this planet's surface. Plate tectonics, too, which requires limestone—made by marine microbes—to lubricate the continental plates, may also be dependent on life. Because a continuous marine rain of algal skeletons ultimately forms a layer of calcium carbonate that lubricates Earth's giant continental plates, life seems involved in plate tectonics. Crashing plates also open up expanses from which ocean water can evaporate, leaving behind salt and thus removing it from the ocean where it can, in too high concentrations, become deadly to most marine life-forms.

Without the continental plates, which move on life-made limestone and require life-recycled water, there might be no Himalayan or Andean mountain ranges. Microbes really do move mountains, and life and the environment are connected with a depth and intimacy that may surpass the imagination of even the boldest scientists.

Convolutedly, the wet recursive chemical system of waterlogged life seen from space as our turquoise blue beauty of an aqua planet might not be here without life. Water is made of hydrogen, but hydrogen, the lightest element, tends to escape into space, where it has gathered again around the giant outer planets. As Stephan Harding of Schumacher College points out, one of the escape routes for hydrogen ever since the Archean eon has been a chemical reaction between water and basalt, which is the major rock type at the bottom of the oceans. Basalt contains ferrous oxide and in the presence of carbon dioxide strongly captures oxygen atoms from seawater, letting hydrogen molecules escape to space. Life is moist and damp, and we live on a happily wet planet, but this natural process could have dried Earth up in two billion years. Bacteria again saved the day: photosynthetic life at the sea's surface liberates free oxygen, which binds to the escaping hydrogen making new water molecules. Other bacteria, such as chemautotrophic anaerobes dwelling in ocean bottom ooze, combine escaping hydrogen with carbonated seawater, creating water in the ocean and liberating methane gas.

Life's alteration of this planetary "rock" is so complete that Earth would look to us to be barren and alien beyond the most hopeless desert if it were not for life and its naturally intelligent, highly evolved metabolic ways. The intuition that Earth looks alive from space has some basis in fact.

Nature is not just red in tooth and claw but quite exuberantly white with apatite, golden with chrysophyte algae, purple with photosynthetic plastids, and peachy orange with carotenoids eaten by long-legged filter-feeding flamingos, whose diet of shrimp and algae and cyanobacteria shows up under their skin. It is butterscotch gold with naturally archiving, insect-preserving amber, beige with *Beggiatoa* bacteria, and black with the magnetic teeth of chitons. It is green with *Chlorobium* sulfur bacteria and red with *Chromatium*, a purple sulfur bacterium that, like its ancient metabolic brethren, oxidizes sulfide rather than water, excreting yellow sulfur rather than oxygen as its photosynthetic waste. Nature is not just dog-eat-dog but bacteria-inside-archaeon, protist mingling with protist, quorum-sensing swarm-sensing crowd-controlling human-making swarm. There is more in it than we dreamed; it comes in colors and keeps on coming, psychedelically without pause or apology.

The night is black and the sky is blue. The sky is blue because of life. It is blue because of oxygen atoms released billions of years ago by green cyanobacteria. Exuberant and dangerous, they spread, poisoning the planet with their toxic oxic pollution. They were the first to be victimized and among the first to be transformed, forced as they were to evolve ways to tolerate and eventually make use of the life-produced reactive gas.

Why is the sky blue, no-longer-jumping Jack asks. A correct answer would be cyanobacteria, the mean lean greens that give summer its verdant sheen. The sky would not be its beautiful aqua or azure, the oceans would not reflect the sky's blue kiss without these frisky germs. For the sky's blueness, its tranquil aerial fields, comes from a surfeit of atoms of oxygen. Their minuteness is a right size to catch a light wavelength and send it spinning and scattering all directions every day. The short, blue-wavelength light reflected by atoms hits our eyes. The early world may have seemed pale and ruddy, and smelled sulfurous and foul by comparison. Inhaling oxygen first released in any meaningful way into our atmosphere by slimy cyanobacteria, we can only say that life on Earth is, if it is, a gas, because these photosynthetic bacteria spread, about two billion years ago, like green wildfire. They directed the energy of light to break the molecular bonds of water, getting a metabolic pathway up on the purple sulfur bacteria that are otherwise much like them. They incorporated hydrogen into their cells and released reactive oxygen into the atmosphere, giving us something to think about, and breathe.

We live on a psychedelic planet where partners triumph, the chemistry of the early solar system lives on in our bodies, individuality appears at higher levels, and evolution evolves. Once upon a time, billions of years ago, Earth's sky was not yet blue. This is because life liberated such oxygen atoms from the hydrogen molecules to which they were molecularly bonded. The size of oxygen atoms is such as to scatter light of a blue wavelength. The life-form responsible for this release of oxygen atoms was cyanobacteria. Their forerunners were purple sulfur bacteria that had thrived on hydrogen sulfide. The new water-using cyanobacteria took their hydrogen directly from the liquid medium in which they lived. Breaking water's molecular bonds was difficult, but it saved early life from perishing and led to a life-form—a very successful life-form—that is arguably the dominant form of life on Earth today. And they were nicely named, considering what they did: cyanobacteria, cyan denoting any of a range of colors in the blue-green range of the spectrum. These blue-green sunlovers became trapped in the ancestors of plant cells, inaugurating the eternal salad days land plants still enjoy. Language itself contains fossils, and thus it is that these beings are still more often referred to by their old name—"blue-green algae"—which tends to obscure what they are: bacteria, symbiotic aquatic oxygenators of a planet that, once upon a time, was a whiter shade of pale. It is an amazing story, all the more powerful for being both mythic and true:

Life gave Earth the blues.

CHAPTER 10

MOUSETRAP

Γνῶθι σεαυτόν
(Know thyself)
—*Oracle of Delphi*

WHY ARE WE HERE? Might this all just be a big fluke? Even if evolution is, as Arthur Koestler said, like an "epic recited by a stutterer,"[1] what is the plot? It seemed like God had a good idea, but then he got sidetracked. Where is he going with this thing?

I believe the writer Kurt Vonnegut touched on the heart of this question. Before a full house of mostly women at Smith College, he first drew a chart that graphed stories. On the X axis he drew time, on the Y happiness. By making a line, he showed, he could map any human story. Goldilocks and the Three Bears started off with a jump when she found the house in the woods, it moved higher like a stock as she saw the table place set for her, then higher again as she found her warm bed, before plummeting when the bears came home. The Garden of Eden started off very high, plummeted down, and then flatlined. Vonnegut used a big sheet of paper to mock scientific reductionism and social science in particular. Most stories weren't so clear-cut or geometric; they were more squiggly.

Then he told about his own days. They often started, he said, despite protests from his wife, who thought he could use his time more wisely, in taking a leisurely walk to the post office to mail a single letter. At the post office he bought a single stamp from the pretty teller. They smiled and he slid her the envelope. Nothing would ever happen, he admitted. But still, that was not his point.

What was it?

"I tell you, we are here on Earth to fart around, and don't let anybody tell you different."

OF COURSE, Vonnegut's is not the last word on the subject. "We are here," writes the paleontologist Stephen Jay Gould, "because one odd group of fishes had a peculiar fin anatomy that could transform into legs for terrestrial creatures; because comets struck the earth and wiped out dinosaurs, thereby giving mammals a chance not otherwise available." Vonnegut's son, the physician Mark Vonnegut, takes a more ethical slant, emphasizing our emotional and physical interconnectedness. "We are here to help each other get through this thing." The poet W. H. Auden was mystifyingly succinct: "We were put here on earth to help others. I'm not sure what others were put here for."

I had a teacher, Don Levine, who taught Avant-Garde Cinema, Madness in Literature, and other fun subjects. One day he informed the class that meaning and moaning share the same root. Although we should be dubious of what Jacques Derrida calls "fabulous etymologies"—words evolve much faster than organisms and are thus even more apt to erase traces of their origins in transit—I would not be surprised. It seems to me that the impulse to do something comes before even the discrete task to be done and is also implicit in the thermodynamic equilibrating of nature, especially the elaborate actions, sometimes including planning, of far-from-equilibrium systems that perish if they lose access to the energy gradients that sustain them. A "forness" or "towardness" blankets life at all levels: physiological, unconscious, and conscious. A couple of days ago, for example, I awoke knowing I had to do something. The unusual part was I did not know which of three things. One, I had jammed my finger while playing basketball two days before and it was throbbing with pain that could be addressed. Two, my bladder was full. Three, I had a sleep mask on, it was Sunday, I had taken melatonin the night before, the mask had fallen off, and I wanted to get it back on before the morning light stung my eyes. What I noticed was that these three desires—treating the pain in my swollen finger, going to the bathroom, and not having light on my eyes—all vied equally for my attention. None was so strong that it obliterated the others, and I initially was not aware of any of them but only of a general desire that something needed to be done. We might call them "telons," from the Greek *telos,* "end" or "goal."

One way or another some members of life have successfully taken care of their goals for the last 3.8 billion years. In a little-known book that has been considered a *Principia* for biology, Robert Rosen identifies anticipation as a key property of life.[2] He argues that its ends and means are entailed, that life is noncomputable and naturally purposeful. Rosen points out that when causes and effects are nested together, purposes become real. I, too, would argue that life's forwardness and towardness comes before any discrete task—any, that is, except for gradient reduction. Life is a cyclical system, so its ends and means must be entailed, but organisms can become their own purposes only within the larger context of gradient destruction in which they are already always engaged.

Try to be completely still. You can't do it. Your heart beats, you breathe, you continue to metabolize. Even sleeping you produce more entropy than a corpse. This, not reproduction, is the essence of life. It is the essence in that it explains not only what we are, aggregates of cosmically abundant matter like hydrogen and carbon, or how we are, but *why* we are, to spread energy more rapidly, more long lastingly than would be the case if we were not here. We are manifestations of the second law, as one of my coauthors, Eric Schneider, has said.

In one of his early books, *The Cosmic Connection,* my father had written words to the effect that the best evidence extraterrestrials might find that Earth has life would be the spectroscopically measurable excess of atmospheric methane, a product of "bovine flatulence."[3] Although there are more important biospheric sources of atmospheric methane than the methanogenic archaea in the rumens of cattle, my father's throwaway phrase is a memorable bit of rhetoric. Whether belching or farting, passing gas lets loose thermodynamic wastes. Compared with the oxygen and food we take in, the carbon dioxide and methane and other carbon compounds we put out are in a less energetic state. They are not completely energetically stable, however, as the microbially aided breakdown products of food include tiny amounts of hydrogen, carbon dioxide, and methane that combine in the large intestine with hydrogen sulfide and ammonia to make a redolent, combustible mixture: Relatively stable, if smelly, these compounds can, if provided activation energy, be caused to burn, as the psycho rockabilly crooner Angry Johnny, the lead singer of the Killbillies, demonstrated one night by nonchalantly taking a lighter to his rear end in Hell, his name for his studio in Easthampton. This created an invisible torch at some distance from his pants.

Methane, although a waste produced by cows, termites, swamps, and others, is still highly unstable in the presence of oxygen, which is why intelligent aliens might notice it, this cosmic, spectroscopic equivalent of a fart. Indeed, here is a solution to the Fermi paradox: the aliens have detected us but don't want to come close for fear of Earth's smell.

The entire genesis of the Gaian idea is that Earth's atmosphere is out of thermodynamic equilibrium, more complex than it should be, given the rules of chemical mixing. But this persistent admixture is explained by the fact that the living organisms of Earth are open systems. Their complexity bleeds over into the environment, even as they produce more entropy (molecular chaos), mostly as heat, farther out. Life is one of a class of thermodynamic systems that naturally organize themselves as they grow by tapping into, and sometimes depleting, local energy stores.

But excess can be bad. It can be fatal, as in death by drug overdose of a Jimi Hendrix, or worse, it can be speciescidal, as in the overpopulation of a species that uses up all its resources and dies of famine or becomes prey to an epidemic. Flatulence and laughter represent "human" aspects of any number of processes that naturally arise to perpetuate the universally observed energy-spreading mandate codified by the second law. Moreover, laughter and farting, although they seem to represent more or less random examples of gradient breakdown, are not really. Unlike, say, spontaneous combustion, or world war, or genocide, they represent *moderate* gradient breakdown; they are examples of stable processes, processes in beings letting out steam, as it were, *gradually,* and thus not, in all likelihood, endangering themselves or the external gradients that sustain them and around which, quite necessarily, their activities are organized.

It is indeed an amusing astrobiological irony that through a smorgasbord of sphincters comes in the aggregate such gas that it provides, within our atmosphere one-fifth oxygen, with which such gases should react, a kind of chemical signature indicating the presence of intelligent life on Earth. Compounds such as the butyl mercaptan released by skunks, methane released by archaea and belching cows and wetlands, and butyric acid released from mammal skin all react strongly with oxygen but linger in our atmosphere—because life keeps farting them around in the shared ether of its collective home.

IN 1948 the Dutch artist M. C. Escher made a famous lithograph. Two hands emerge out of sleeves on a piece of paper to draw each other. Ends and means become beguilingly intercalated in highly entangled complex systems, tending to obscure the basic operations of those systems, especially when they themselves possess semiotic and interpretive abilities. In the 1790s Immanuel Kant said that the difference between an organism and a watch was that an organism creates all its own parts. An organism, unlike a watch, is "both cause and effect of itself." Kant, who meditated on the difference between a watch, a mechanism that is made, and an organism, a being that grows, was one of the first to think about self-organization. But he was, I think, a little off, as are those who follow on that train, because the organism is connected to an environment and, indeed, dependent on it and the continuous telic reduction of its energy gradient for its own continued sustenance. The internal nesting of causes and effects obscures a function, or purpose, which becomes far more clear when one takes into account the environment. Kant also described an organism as its own purpose. But there is no self disconnected from the material substrate and energetic flows on which it depends. It only ever cycles in order to reduce ambient gradients.

Kant, aware that religious humanity tended to think of itself as made, like a machine, for a higher being, also focused our attention on the difficulty we have determining purpose. The nestedness of cause and effect in cyclical systems disturbs linear logic in interesting ways. For example, as a cyclical system I may be hungry. Therefore I eat a cupcake. Therefore I am not hungry. If we remove the intermediary behavior, we end up with the logical contradiction, "I am hungry . . . therefore I am not hungry." Still, it makes perfect sense within the cyclical system of which it is a part.

The cells responsible for clotting blood near a paper cut on your finger, if they were conscious, would not necessarily have a clue what they were up to participating in a cascade of enzymes that stops blood flow and heals the damaged tissue, repairing the wound. The purpose becomes evident only as we move outward to see how the organism functions, not only in terms of itself, which it does do, but also in terms of the environment.

Without knowing, or being familiar with, certain details of the cyclical operation of a complex system, we may completely misrecognize or overanalyze its simple natural function. A story by the great French writer Denis Diderot (1713–1784) beautifully illustrates the point. This story is of a wayward horse ridden by the servant Jacques in the novel *Jacques le fataliste et son maître* (Jacques the fatalist and his master). Jacques, who behaves rebelliously, willfully, even though he believes in fate, as if everything were already written above in the stars, takes orders from his nobleman, who, ironically, although he believes in free will, is more happy-go-lucky, rarely taking any initiative. In any case, Jacques one day is on a horse that appears quite willful and takes him, Jacques, off the beaten path and certainly not where Jacques has been telling him to go. And where he ends up is indeed fateful. The horse, moreover, as if making sure Jacques understands, leads him there not once but twice.

And where does he lead him? The gallows.

Naturally, Jacques believes that the horse is sending some kind of message. And the message is clear; it is one of impending doom. For whom, Jacques doesn't know. But he cannot help but think that the horse's repeated purposeful behavior is a sign that the grim reaper may be on Jacques's tail.

The third time the horse goes off course, Jacques arrives at the property of its previous owner, the executioner. There was nothing sinister in the horse's purposeful movements: it was simply returning to places it knew well, places where it was used to being cared for and fed.

If Diderot is arguing that the horse's true purpose is not as ornate as Jacques superstitiously imagined, I am arguing that neither may be ours. We have our little purposes and indeed big purposes, such as finding meaning in life, a purpose that gives us the will to live. But there may be bigger purposes. Indeed, if our immediate purpose is to get a bite to eat or a date, to make money or a baby, or even to raise a family or win a war, if we step back we may be able to see that these take place in a larger frame. They take place in the cosmic context of a naturally telic, purposive universe.

MY LIMITED RESEARCH SUGGESTS that Vonnegut's humorous and partial nonanswer to why we're here was not just a throwaway line. He gave it some thought. It was the highlight not just of his lecture but, in a more extended way, of the plot of his novel *Galapagos.* Here he tells the story of rich tourists stranded on an island after the rest of humanity has perished from world war and epidemic. The narrator, the ghost of a shipwrecked sailor who drowned but now speaks from the pink tunnel of the afterlife, tells of how the last old man, feeling responsible for humanity's future, among other things, takes matters into his own hands, literally, introducing semen into the young women in their slumber to preserve the species. The result is healthy young who possess some odd traits. They evolve hairless and develop an extra layer of blubber. After a few generations, their heads are so pointy that they enter the water without a splash. In a few more, none of them remember their mothers after the age of three. Finally, their only trait that ties them to the species from which they came is that whenever they lie about in the sun and one of them farts, the rest laugh.

Combining the ethical precept that we are to be judged by what we do, not what we say, and the materialistic perspective of modern science that brackets intentions, looks for physical causes, and weighs material evidence, I did a Vonnegutian calculation of my own. I figured that the average human being produces roughly thirteen tons of shit over a lifetime. That's going too far no doubt, but it could be used to argue, since 99.7 percent of "our" total weight is poop, that that is our purpose.

I am aware of the *post hoc ergo propter hoc* fallacy that just because something happens after something else does not mean that it was caused by it. But it is not that simple in a cyclical system, which we are. The odorless methane and smellier hydrocarbons of flatulence are thermodynamic waste gases. They are produced by cyclical thermodynamic systems during their ongoing operations. These operations, as rarefied and complex as they may have become, continue to perform the basic telic task of the second law. What is humorous about farting is not necessarily its inessentiality as is its status as a sign of our essential nature as thermodynamic systems. All of the living ones produce waste gases. Using solar and chemical energy, bacteria cycle these gases globally. We do not. We are a very partial, incomplete part of this system, and farting secretly alerts us to this outside, our insufficiency within the whole. We exclude what we exclude, but it comes back as a wake-up call: We are missing something, the essence of our nature, which is to reduce gradients and preserve ourselves as gradient-reducing beings. Our telonic nature comes before all discrete tasks. So I would say that Vonnegut, though he was joking, was basically right. We are here to fart around. It is funny because it's true.

IN *THE BIOLOGIST'S MISTRESS* the novelist Tori Alexander writes of how, having converted to atheism in college and having become an avid Darwinist, she would cross out teleological language in her textbooks; for example, she would change "Birds have feathers *in order* to keep them warm" to "Birds have feathers *that happen* to keep them warm."[4] Replacing purpose with chance in this way removed illicit, outmoded connotations of a hands-on deity unscientifically meddling with lawful nature. Since she grew up in the Bible Belt, she considered rooting out purpose, associated with religious superstition, to be part of her education. Later, however, when she became a novelist, she realized that purposefulness is fundamental to living systems, to the structure of art forms as well as organisms. Artists and organisms incorporate random factors, but they do not do so randomly, but purposefully. Such purposefulness, however, is not theological.

As Stanley N. Salthe, a self-described "apostate from Darwinian theory [who has] described it as part of modernism's origination myth,"[5] points out, thermodynamics provides a level of final causation that is consistent with a natural law-governed cosmos, a source of purpose that is neither theistic nor evolutionary. Fidgeting such as tapping your foot serves no higher purpose but adds to gradient reduction. Lightning reduces an electrical energy gradient, thought the redox gradient of the blood sugar in your brain. But these are not the higher forms of purpose we usually associate with a plotting God. They are more like necessary activities in a universe with a cosmic energy mandate. Life shares traits with nonliving systems that maintain complexity and grow *to* accommodate energy's tendency to spread. It is a natural form of energy transformation in a thermodynamic universe.

Still, however, the fastest ones—and here we may consider ourselves, in our cosmic context as an energy-finding, energy-degrading system—may compromise their own effectiveness. As an energy theorist and pioneer ecologist, the mathematical polymath and insurance adjuster Alfred Lotka suggested, life seems more to optimize than maximize energy use. In a sort of trial run for what Sigmund Freud would later describe as the reality principle, life has learned, no doubt the hard way, to moderate the taste for resource destruction that nonetheless fuels it.

Entropy production is something that happens anyway—it is codified as the second law of thermodynamics—but complex systems are better at it. They do it more efficiently—more quickly, more long lastingly, or both. From a view that the ghost of Vonnegut may share, there is no divine purpose in the sense of consecrating human beings as a "chosen" species. But there is, as his evolutionary thought experiment, *Galapagos*, shows, a little purpose, one that, at least in Vonnegut's capable narrative, both we and our imaginary progeny share. From this admittedly evolutionarily reductionistic perspective, the big difference between us and them is that the big-brained species grew too fast and thus blew its chance, whereas the better swimmers kept things within limits, not becoming too big for their ecological britches. Along with Γνῶθι σεαυτόν, "Know yourself," Μηδὲν ἄγαν—"Nothing in excess"—was inscribed in gold at the Oracle of Delphi, near the base of Mount Parnassus in the Valley of Phocis, near a rock called Omphalos, regarded as the center of wisdom in the classical world.

IF YOU'VE EVER SEEN the Milton Bradley game of Mousetrap, you've seen how an initial impetus can lead to a cascade: like a line of dominoes, an initial impetus causes a chain reaction, leading to follow-on actions until the energy is dissipated. Each domino's fall is caused by the previous domino. Life is different in that the dominoes never stop falling. Life's trick has been to catch energy and not just send it spinning in a captivating series of Mousetrap-like secondary reactions but—beautifully, amazingly, stupendously—to store some of the energy that it takes in and then, later, exploit that stored inner energy to go out and find another finger (energy source) to push the first domino.

The idea of purpose is usually avoided in science. It seems to imply a mystic reverse causation that cannot be scientific. The reticence is that effects in the material world flow from a prior cause to a future event, while effects in the realm of purpose, which seem to belong only to the realm of the mind, explain present activities in terms of goals not yet achieved. We don't mind attributing such behavior to us, because that's how it feels, but we are stingy to see its existence in others, even the "higher" animals (e.g., similar to us). But the truth is, purposive behavior is everywhere in nature and doesn't require brains. An amoeba swims toward nutrients, a bacterium toward sulfate or oxygen. Sunflowers follow the sun.

A cue ball slamming into an eight ball sends it flying, mechanical cause and effect. But your hand reaching for a key is a different story. The key was lifted because it was your purpose to get it to open the door. This is a very different kind of operation, outside the usual scientific realm of immediate cause and effect. Indeed, many people would think that intention, in the form of wanting something for the future, then achieving it through an act of will, is something that happens only to people and some animals, and God.

The truth is, we act as loops in which past and future are intimately conjoined. We know from our experience the familiar feeling of desiring something and doing it, satisfying a drive. There is nothing mystical about such causation in life.

In Mousetrap, a ball bearing, sent rolling, zigzags down a blue ramp into a red slide. There it trips a yellow lever. Which releases another ball bearing that falls through a miniature plastic bathtub onto a toy blue diving board. This seesaws a green plastic diver into a yellow cup. The cup, vibrating, jostles a yellow pole. Which in turn releases a red mesh bell—the final goal of the whimsical contraption—the fall of the mousetrap.

Why do we find this series of connected actions so intriguing—spellbinding in a way similar to a child arranging and then watching a Slinky curl down the stairs? The reason is that it is kinetically familiar. It is a setup in which a series of actions come together surprisingly to give unity to mechanical behaviors. Energy is gathered and transformed, and its release triggers another phase in an unfolding routine.

Life's animation is similar. It is as serendipitous as the stair-walking Slinky, as intricately connected as the colorful toys of Mousetrap. Like them, life stores and redeploys energy in a series of connected actions.

But the "goal" of life appears not to be to catch a plastic mouse or arrive at the foot of the stairs. Life is like a Slinky, but its parts are connected chemically rather than mechanically, and the final result is not a modulated drop down a discrete number of stairs but continuous energy expenditure via structures that release energy to build new, similar structures.

You are one of these structures.

PART IV

CLOSING THE OPEN CIRCUIT

CHAPTER 11

PRIESTS OF THE MODERN AGE

Scientific Revolutions and the Kook–Critic Continuum, Being a Play of Crackpots, Skeptics, Conformists, and the Curious

One should as a rule respect public opinion in so far as is necessary to avoid starvation and to keep out of prison, but anything that goes beyond this is voluntary submission to an unnecessary tyranny, and is likely to interfere with happiness in all kinds of ways.
—*Bertrand Russell,* Conquest of Happiness

People nowadays no longer believe in originality of single people and small groups. Everybody believes in the big group and in the joint power. We have a Maoism in science. Let flowers grow. It's no longer likely to happen. Everybody believes the ideology that it's no longer possible to be a Poincaré or an Einstein. But we also live in the Age of everybody believing in the Big Bang, which is the greatest nonsense of all, if my co-workers are right. And yet it's impossible to get rid of it. We live in a dogmatic age. People want to derive certainty from common opinions. They don't believe it's possible to find something really original. It's a pity for our young people. They're not allowed to believe in themselves anymore.
—*Otto Rössler, "Interview: Professor Otto Rössler Talks on the Large Hadron Collider (LHC)"*

The moral of the tale is the power of reason, its decisive influence on the life of humanity. The great conquerors from Caesar to Napoleon influenced profoundly the lives of subsequent generations. But the total effect of this

> influence shrinks to insignificance, if compared to the entire transformation of human habits and human mentality produced by the long line of men of thought from Thales to the present day, men individually powerless, but ultimately the rulers of the world.
> —*Alfred North Whitehead,* Science and the Modern World

WALLS OF ORTHODOXY – AND THE REBELS WHO BREAK THROUGH THEM

"Scientists are the priests of the modern age, and they must be watched very closely," wrote Samuel Butler at the end of the nineteenth century. Butler had converted to an evolutionary view after he read Charles Darwin's *Origin of Species.* Since Butler had freed himself, with great difficulty, from his father's religious doctrine and its ambience, Victorian hypocrisy, he refused subjugation of his critical, curious mind to yet a new authority.

If Giordano Bruno had been burned at the stake and Galileo Galilei put under house arrest for following their open minds and engaging the evidence—threatening the ecclesiastical arbiters of truth in the form of religious doctrines—so the rise of science as a thought-style threatened to erect a new repressive social structure, one ironically created by science's institutionalization of open-mindedness and error correction in the form of the scientific method. Butler's comments were provoked both by his great intellectual excitement upon reading *The Origin of Species*—he was convinced—and his disenchantment (after reading some of the predecessors Darwin himself had acknowledged) with Darwin's too-mechanical presentation. In treating organisms as objects created by natural selection presented as a Newtonian-like law, Darwin had removed one of the most important characteristics of organisms themselves from his description: their agency and autonomy, their self-originating ability to alter themselves and their environment. The New Zealand–based Butler became further disenchanted when he tried to contact his former neighbor and was given the cold shoulder by both Darwin and Thomas Huxley, his articulate champion and "bulldog." Although Butler wasn't a scientist, he accused Darwin of taking the life out of biology, and stonewalling dissent as a new authoritarian social structure began to arise.

Butler's statement of scientists as priests may surprise some, especially in the United States where religion and science, especially evolutionary biology, are often seen to be at each other's throats. But his comment was sociological. As an iconoclastic freethinker who had resisted his family's attempts to push him into the clergy, who had stopped saying prayers, and who had fearlessly explored Victorian society's hypocrisies and pretensions, Butler was acutely aware religion had no monopoly on dogma. The closing of minds around received wisdom, even ideas that were rebellious in their own days, seemed a natural process, like hardening of the arteries. Dogma was mercenary, and already metastasizing to the culture founded by those who stood up to religion. It is no coincidence that, as Stephen Jay Gould pointed out, both Galileo and Darwin published important scientific treatises in a popularly accessible form. Galileo's *Dialogue Concerning the Two Chief World Systems* made the case for heliocentrism in a popular tract that put words in the mouth of a Galilean detractor named Simplicio, and was published in Italian, the language of the common people, at a time when serious scientific pronouncements were as a matter of course published in Latin, the cultured, would-be universal language of both science and the church. In 1633, based on this popular science book, Galileo was placed under "grave suspicion of heresy." Not only was the *Dialogue* placed on the Index of Forbidden Books (and not taken off until 1835), but a secret edict agreed to ban anything else Galileo had written or would write.

Although not as flamboyantly, Darwin also stood up to the church. He had gathered an immense amount of material that refuted the church's contention that species were created once and for all. Darwin could have aimed his book at specialists but instead published it in common language as a trade book. In both cases, Gould argued, the main ideas—for Darwin, evolution by natural selection, for Galileo, Nicolaus Copernicus's counterintuitive Sun-centered solar system—were too important to be adjudicated by ossified institutions of knowledge. Sometimes science is too important to be left to scientists: here we see two world-class scientists implicitly agreeing that the public must directly be informed, that they have the right to vet the ideas themselves. Darwin was more cagey than Galileo—he mentioned a Creator at the end of *The Origin*[1]—and censured Galileo was more politically savvy (or cautious) than poor Bruno, but both chipped away at the absolute authority of the church, assaulting faith with facts and introducing falsifiability in place of infallibility. And as the star of science, in the hands of such brave and savvy men, rose, so did their own authority and that of science as an institution. Today of course, for many if not most people in the world, science is granted greater powers to peer into the truth of things than religion, which skeptics and atheists regard as an atavism, its texts preserving contradictory doctrines, superstitions, and knowledge current thousands of years ago that has been buttressed by faith and repetition rather than being put to the test.

Still, belief systems and the authority of those who propound them are irreducibly human. As an institution, science, like religion, is subject to the quirks of hierarchical social relationships. Laypeople—we retain the term from those not in the clergy—must trust in the seemingly far more knowledgeable pronouncements of those in the knowledge business. (Science comes from *scientia,* Latin for "knowledge.")

One might argue that the more radical a new theory is, the less likely it is to be accepted by established authority and the more likely it is to offend those who've invested their careers in a threatened scenario. This essentially conservative dynamic stymies pseudoscience and crackpottery but also presents hurdles for those with more correct, or useful, ideas who come up against the prevailing authoritarian intellectual power structures. More daring than Darwin, Galileo himself was skeptical of the groupthink not only of religionists but of scientists themselves, who were by no means immune to ignoring evidence and reining in originality and testing in the face of consensus. "Assignment of the Science is not to open a door to endless knowledge, but to set a barrier to the endless ignorance," he wrote. For Galileo, a lone thinker in his basement could be more productive, get closer to truth, than a group of pontiffs or experts whose judgment was clouded by peer pressure. "That man," he writes in *The Assayer* (1623), "will be very fortunate who led by some unusual inner light, shall be able to turn from the dark and confused labyrinths within which he might have gone forever wandering with the crowd and becoming ever more entangled. Therefore, in the matter of philosophy, I consider it not very sound to judge a man's opinion by the number of his followers."[2] The philosopher Friedrich Nietzsche espoused a similar sentiment, to wit, that madness in individuals was rare but common in crowds.

Science, unlike religious dogmatism, is supposed to be methodologically open to novelty, welcoming new perspectives if they are supported by the evidence. Unlike the ecclesiastical authorities, disparaging our demotion from the center of the solar system, Galileo was very excited not only about the more realistic, new view but about change in general. No longer at the center, Earth became part of a "dance" of changing celestial objects. Whereas the old view of a central Earth and fixed spheres of stars offered the solace of stability and confirmation of biblical and Aristotelian astronomy, the new view, based on telescopic observations, showed a universe full of surprises that included Earth and its organisms in a kind of cosmic community. Change, even if it threatened the religious doctrine of eternal life, was to be welcomed, not feared. "If the Earth were not subject to any change," wrote the astronomer in the 1632 *Dialogue*, "I would consider the Earth a big but useless body in the universe, paralyzed . . . superfluous and unnatural. Those who so exalt incorruptibility, unchangeability and the like, are, I think, reduced to saying such things both because of the inordinate desire they have to live for a long time and because of the terror they have of death. . . . They do not realize that if men were immortal, they would never have come into the world."[3]

LIVING *IN* HISTORY

The practice of science must be distinguished from thinking for oneself that is the spirit of science. We tend to behave as if we live outside history, but we live in it: many ideas accepted by our culture today will be considered laughably incorrect or even tragically wrong by our descendants. Sometimes ideas that seem ludicrous turn out to be true; sometimes ideas that we have rightly dismissed as being insufficiently critical turn out to be true in another, deeper sense. Consider the disproof of the spontaneous generation of life by Lazzaro Spallanzani and Louis Pasteur, true enough now but not possible if life is believed, as many working scientists speculate, to have originated on Earth. Or again, take heliocentrism: yes, Earth travels around the Sun, but this deep and founding truth (the sense of "revolution" in scientific revolution derives from Copernicus's realization that Earth revolves) is nonetheless compromised by relativity theory's assertion that there is no privileged frame of reference. Thus, in a real sense, we can regard the Sun as revolving around Earth after all. Still another example is Empedocles's notion of evolution by natural selection and symbiosis (albeit of organs) on the ancient Earth: both correct but premature theses were dismissed by Aristotle, who probably thought they smacked of the wild stories of sex beyond species borders familiar from Greek myths. (Aristotle, who had enough to deal with trying to reverse his teacher Plato's emphasis on an unseen unchanging world, may also not have liked Empedocles's notions because they flouted Platonic formlike species boundaries.)

TRUTH IS NOT ABSOLUTE but infiltrated by a changing consensus. This is an idea advanced within social studies of science and is a conclusion of Peircian pragmatism. That agreed-on truths are subject to intellectual fashion is hardly contestable. In an example of epistemological nihilism, the literary agent John Brockman, who today represents the most successful scientist–writers but who once helped organize events with the iconoclastic artist Andy Warhol, once wrote, "Nobody knows and you can't find out." Such a sentiment is the antithesis of a naive faith in a scientific truth removed from human mediation, and I think overstates the case. "Reality is that which, when you stop believing in it, doesn't go away," wrote the science fiction author Philip K. Dick. The famous economist Joseph Schumpeter said something very similar, that the quality of truth is that it exists whether or not you believe in it. What is science's ideal? To quote the quantum physicist David Bohm (1917–1992), "Science is the search for truth whether we like it or not."[4]

That wishing won't make it so is common sense, but common sense, when it challenges widespread beliefs (which, as articles of faith, do not require active thinking), is not so common. And if one engages critical thinking but goes against the multitude, one can be ridiculed, ostracized, or even killed.

When—to cruise backward in time to briefly show a few salient examples of this phenomenon—James Shapiro wrote a popular article in *Huffington Post* summarizing his *Evolution* chronicling the new evidence of transposons (jumping genes), redundant DNA, regulatory RNA, and other discoveries of modern molecular biology that suggest the genome has an ability to rapidly reorganize itself—contradicting the neo-Darwinian tenet of gradual evolution by random mutations—he was lambasted by a University of Chicago colleague on his Why Evolution Is True blog as "completely wrong" and "misguided."[5] And it was suggested, in the largely anonymous comments section, that if he wasn't a creationist he was playing into creationist hands.[6] When I communicated with Shapiro, he assured me that he was not a creationist and that of course there was a role for natural selection, but it should not be used as a "deus ex machina" to explain all complexity—and that empirical investigations showed that evolutionary novelty came from symbiosis, hybridizations, and internal genome restructuring. Neo-Darwinism (whose adherents refer to their beliefs more generally as "modern evolutionary biology"), he characterized, ironically considering its adherents' quickness to dismiss critics as creationists, as a "religion" for its reticence to honestly consider evidence and alternative views.

After Case Western Reserve University withdrew its press release that its assistant professor of molecular biology and microbiology, Erik D. Andrulis, had proposed an ambitious new "experimentally and experientially verifiable" theory of life claiming to "unify quantum and celestial mechanics" around a geometric entity he called a "gyromodel," he was accused of "mental illness" and "Sokalism"—the latter being a reference to a famous practical joke the physicist Alan Sokal played in getting a jargon-infused sham paper accepted by the cultural studies journal *Social Text*.[7]

When Otto Rössler, a German biochemist and author of three hundred papers on chaos theory, warned that CERN has no safety guidelines to protect against a one-in-six chance of particle collisions creating possibly Earth-destroying black holes, his fears (and the lawsuits addressing them) were dismissed.

Now we live under a sort of temporal distortion that one might call the Logarithmic Illusion of the Now: the past seems settled whereas the present seems relatively magnified, swollen, and urgent. The examples above are relatively unsettled. How do we distinguish between crackpots and curious seekers with good criticisms or alternative theories that need to be investigated in the moment? There is no magic sword, especially not one sharpened with the whetstone of orthodoxy, that will magically separate science from "pseudoscience." Without assuming that past vindications prove anything about the present, it is obvious that ridicule and claims of madness are not equivalent to valid critique.

When the Italian inventor Guglielmo Marconi (1874–1931) claimed he communicated through the air without wires, some of his acquaintances were convinced enough of his craziness to suggest he be put in a mental institution[8]—yet just today you may well have checked your e-mail, spoken on a cell phone, or be reading this online.

THE RISE OF OUR MODERN SCIENCE "thought-style" now threatens Brunian–Galilean young open minds with new repressive social structures.[9] Ironically, despite science's touting of open-minded error correction as the key to "the scientific method"—as if this were only a single procedure—new kinds of repression have been created by science's institutionalization.

It is always safer to say what *can* happen than to say what can't based on conceptual first principles or theoretical prejudices. After miscalculating the age of the Sun, the physicist William Thomson, or Lord Kelvin (1824–1907), from whose name we get the Kelvin scale in temperature, argued that "heavier than air" flying machines were impossible, that X-rays were a hoax, and that radio was a doomed idea. The first of the three "laws of science" coined by Arthur Clarke (1917–2008), the science fiction writer and proposer of the idea of a telecommunications satellite, is that, "When a distinguished but elderly scientist states that something is possible, he is almost certainly right. When he states that something is impossible, he is very probably wrong."

The correct conclusion to be derived from the attempt to have Marconi committed to a mental institution is not that we inhabit the insane universe of Marconi's demented imagination but that the masses of his contemporaries, including some very smart people, were dead wrong. Although the incessant questioning and critical judgment of organized science led to wholesale revisions of humanity's views of reality, it arguably did not change human nature—hierarchical, gullible, and authoritarian, with a perhaps evolutionary-based weakness, certainly affecting scientists themselves, for people to attribute correctness to their own beliefs.

TOO IMPORTANT TO BE LEFT JUST TO SCIENTISTS

Openness to new ideas threatens to depose the authority of the masters of the old ones, even and perhaps especially, one might argue, the scientists who have staked their methodology on the welcoming of novelty if merited by the evidence; this is because views arrived at after open debate and testing, rather than faith and suppression, enjoy greater legitimacy and efficacy and thus seem, paradoxically, still more impervious to change. Scientists are the priests of the modern age: As science becomes more institutionalized, more powerful, Butler's comment becomes more poignant, more profound. To be fair, it seems that scientists personally deserve our vigilance less than the institutional structures that fund science, and of which scientists are a part. For me, Butler's salient comment has also lately taken on a personal aspect. Like Butler, I am not a scientist, but like him I believe that the scientific attitude comes out of deeper philosophical stance of curiosity and critical thinking, a stance that must face dogma even at the risk of social disapprobation. In a media age, in a public relations state where science can and has been corrupted by corporate bottom lines and government-sponsored agendas, the philosophical heart of science—thinking things through for yourself, and seeking the truth whether we like what we find or not—is increasingly in peril. I was recently exposed to an extreme example of scientific groupthink. The German intellectual Walter Benjamin (1892–1940), who died before his time fleeing the Nazis, wrote that extremes can sometimes be more illustrative of a phenomenon than typical examples. I think this is the case here.

Let me preface this by saying that non-Darwinians, or better, insufficiently virulent Darwinists, have often been unceremoniously attacked for, for lack of a better term, their lack of faith. Many examples could be adduced, including among Darwinists themselves. Alfred Wallace, whose early understanding of natural selection led Darwin to hurry to publish his similar findings, criticized the notion of sexual selection of bright colors, on the basis that the inside of bodies are brightly colored but not perceived by animals or therefore sexually selected. Gould criticized the tendency to assume all characters have been actively selected for as "panadaptationism." But this hardly makes him anti-Darwinian. And the neo-Darwinian tendency to overextend the explanatory principle of natural selection is well taken. Male nipples (as Gould pointed out) were not directly selected for but result from the embryological similarity of males and females, which of course as mammals require breasts.

Clearly part of the vituperation of those insufficiently genuflectory of natural selection is that, in a polarized political climate where Darwinism is closer to experimental science and creationism closer to the tribal atavism of a prescientific superstition, there is no room for nuanced pastels beyond the Zoroastrian palette of black and white. But "Darwin's dangerous idea" (as Daniel Dennett tagged it—although it was in fact advanced by others, including Darwin's own grandfather, Erasmus Darwin, before him) does not provide a complete explanation for complexity, biological or otherwise. In his memoirs Vladimir Nabokov discusses a butterfly whose wing design seems not only to contain a simulacrum of a water droplet but to accurately display the refraction of the line of a wing marking through it.

His wording as well as his critique of the twentieth-century canon—Sigmund Freud and Karl Marx and to a lesser extent Charles Darwin—makes the reader wonder whether Nabokov is intimating an otherworldy designer rather than naturally selected camouflage. In fact, he may just be calling attention to the human predilection for finding patterns—a kind of conspiracy theory writ small—whether they be those of grand divine design or of a more measured evolutionary sort. The wing marking may not be naturally selected or indeed anything outside the human beholder. This point was brought home to me when I saw a picture book compiled by someone who had found all twenty-six letters of the alphabet reasonably produced by nature on the wings of butterflies. In what I am about to discuss it absolutely needs to be pointed out that anti-neo-Darwinism is in no way necessarily anti-Darwinian and in fact may be more respectful of the cagey close observer, more fond of observation and qualified prose than rhetoric and polemics, Charles Darwin. It is also worth pointing out that Darwin himself could, if one were so inclined, reasonably be called a Lamarckian because of his adherence to a theory of gemmules, theorized agents of inheritance that could pass on changes accrued within the parent' own lifetime. The situation is not unlike Oscar Wilde's barb, "Christ save us from the Christians." Because the stakes of the science–religion battle were set by the debate between evolution and Christianity, even avowedly secular and demonstrably scientific alternative explanations of complexity have been lambasted if they are perceived to counter the neo-Darwinist creed with its emotional investment and faith in the shaping powers of natural selection. In the "extreme" case I am about to discuss, my mother, Lynn Margulis, argued that natural selection was not a source of variation and innovation but a shaper of it. Symbiogenesis is richly supported with demonstrable examples showing its direct role in speciation,[10] whereas speciation by the gradual accumulation of inherited variations, while it claims many more adherents, and is certainly reasonable in principle, has no proven examples. Thus, while the neo-Darwinists doubtless would not want to be added to that Christian proto-existentialist's company, Søren Kierkegaard's argument that the less evidence there is for a phenomenon, the more one must put faith (and one might add, intimidating stridency) in it, applies here.

What happened is that the University of Chicago evolutionary geneticist Jerry Coyne, upset by a recent interview with my mother in *Discover* magazine, lambasted her for being "extremely dumb," "diss[ing] evolutionary biology," "embarrass[ing]" the field and herself, and so on. Coyne said he had been asked by the *New York Times* to review our *Acquiring Genomes: A New Theory of the Origin of Species,* but, though he likes reviewing for the *Times,* he found the book's ideas "so dreadful" that he refused on the grounds that he didn't want to give it any attention. Coyne, who's written the best seller *Why Evolution Is True,* providing ammunition against shortsighted creationist interpretations of the phylogenetic evidence, was particularly incensed by my mother's comparison of neo-Darwinists themselves to "a minor twentieth-century religious sect within the sprawling religious persuasion of Anglo-Saxon biology."

Many of the people who posted on Coyne's website, hosted by the Richard Dawkins Foundation for Science and Reason, launched into a litany of ad hominem arguments, calling my mother a "moron"; suggesting that her ex-husband, my scientist father, Carl Sagan, would be outraged or divorce her; describing her brand of evolution as feminist; accusing her of being a careerist, a narcissist coveting attention, a mischaracterizer of her reported conversation with the evolutionary biologist Richard Lewontin (in which the Harvard biologist said that many neo-Darwinist models were unsuccessful and used mistaken assumptions but were useful because they received funding from granting agencies); and commenters added on to Coyne's proposed theory that successful scientists—especially ones who have proposed controversial ideas before and then been proved right—are captivated by fame in later life into believing their own crackpot ideas, seduced by success and forgoing the laboratory work that got them to their positions of authority in the first place. If Coyne had set the tone with his claim that she was "dogmatic, willfully ignorant, and intellectually dishonest," some readers were quick to join in the largely anonymous bashing.[11]

One of the only two or three commenters who dared to dissent from the prevailing mood of derision and name-calling complained that, after visiting the page a week later, he felt physically sick; he compared the dynamic to that of a "feeding frenzy" as shown on a nature documentary. It was at this point that the website's moderator, Coyne himself, lamented, "Oh, give me a break! The woman made extremely stupid and erroneous comments about evolution and we called her on it." I was particularly struck by one anonymous (in the sense of having just initials as an Internet handle, rather than an accountable real name) comment that pointed out Margulis not only questioned the connection between HIV and AIDs (such questioning being described by Coyne as "dreadful," without further comment) but was a "9/11 troofer [*sic*]" who "pals around with Holocaust deniers, despite being Jewish." There were also many comments deriding her for her role in contributing to the "Gaia hypothesis" of James E. Lovelock, FRS—inventor of the electron capture device, the microwave oven (although he didn't patent it), and an early innovator of cryogenics technologies, among many other scientific contributions.

The Gaia hypothesis in its scientific formulation proposes that our biosphere has modulated its atmospheric chemistry, mean temperature, oceanic salinity and pH, and other variables for hundreds of millions of years at a time because of the interaction of gas-exchanging, environment-altering organisms at Earth's surface. Main tenets of this argument for multiple feedbacks between living and nonliving components in the biosphere have been integrated into geology and Earth system science departments around the world—they do tend to stay away from the word *Gaia* because it is that of the Greek goddess of the Earth and has been appropriated by new age and feminist environmental groups.

Agreeing with the man who compared the ill-mannered blog and its posts to a "feeding frenzy" and intellectual "street mugging," I added and attempted to post my own somewhat acerbic but I think fair comment, which reads as follows: "This blog is an embarrassment. Earth system science is Gaia theory, Margulis in no way denies natural selection, there is a serious question about the relationship of HIV to AIDS in the disease industrial complex, and there is an organization of over 1600 engineers and architects who question the government's explanation for the three buildings that imploded on 911. LM often overstates her case but I have found that she not only has the courage of her convictions but (as Nietzsche said) the courage to go against them when evidence warrants. (By the way in Oxford Richard Dawkins criticized her for not being symbiotic 'enough.') Seriously guys and gals, you can do better—this blog combines the critical acumen of a reality TV show with the emotional tenor of a witch hunt. Why not engage the facts rather than the personalities? Maybe this will help correct the misapprehension about Acquiring Genomes:"—and here I provided a link to my own Amazon review, "coauthorial critique," addressing what the book did and didn't say and providing further, postpublication evidence in support of its main thesis.[12] Perhaps there was some technological glitch, or perhaps the comment was censored, but as of now my comment has not shown up.

Ever since Darwin, science's star has been on the rise. Matters of religion, of ultimate reality as it were, have increasingly come to be seen as amenable to investigation and thus within science's sweeping theoretic purview. This increase in scientific authority has been accepted even by religion itself. Most denominations of Protestantism, for example, now officially accept evolution, as does the Catholic Church, which also endorses modern cosmology's idea of the Genesis-reminiscent big bang as well of course as Earth's being round and in orbit around a star that is one of many billions. It seems surreal to contemplate that an early teacher of Copernicus's heliocentric theory—author of *Dell infinito, universo e mondi* (Of infinity, the universe, and the world)—could be lured to Rome with the offer of a job and then turned over to the Inquisition who kept him chained up at the Castel Sant'Angelo for eight years awaiting his trial. Upon receiving his death sentence and telling the church it was more fearful than he, his jaw clamped shut, an iron spike was driven through Bruno's tongue, and another spike was inserted into his palate.

After being forced through the streets of Rome, the former Dominican friar was burned naked at the stake on February 17, 1600. Some report that the public threatened a boycott of such entertainments if heretics were so silenced, as the nonelective spike surgery had prevented the public from hearing their screams.

We may have come a long way, but we're not there yet. As science becomes more specialized, as scientists become more powerful, and as their work and theories become more embedded in state apparatuses of corporate and government command and control, often requiring funding through universities run as businesses, human tendencies for dogmatism and hierarchy, if not burning at the stake, reappear at another level. Part of the rise of Protestantism was its protest against entrenched authority. The full beauty and truth of the universe cannot be found in a canonical text or by speaking to an imaginary person in the sky. Science is both an intellectual adventure and a spiritual experience. As the Protestants dispensed with priests to show that the individual could have a personal relationship with God, so science shows that anybody on the planet, employing the nondenominational method of science, can have a personal understanding of the cosmos. Science offers that potential, but the primary texts, if not in Latin, become daunting, mathematically and terminologically prohibitive to the average Joe. Still, we want to know, and turn to science writers, to journalists, and to scientists themselves to inform us on the latest and greatest ideas of our global and cosmic reality. Many people still turn avidly to the daily newspaper, although much of it is only gossip. As Stewart Brand has said, science is the only news. Who, what, when, how, why are we? What could be more thrilling? Yet who, as Plato said in *The Republic*, will watch the watchers? The spirit of science is one of questioning authority, putting ideas to the test, and rejecting them if they come up short. But as science assumes the status of cultural orthodoxy once reserved for priests and pontiffs, the stakes, so to speak, become greater, and the individual becomes increasingly ostracized from knowledge formation. The problem for science writers is especially acute, for, although they are the go-to sources for laypeople, they are not themselves qualified to question the new intellectual authorities. Their job has been described as Aaronic, in the sense of just conveying what authorities say, making it plain but not altering it. It is not their place to challenge, but only to spread the word.

Plato's watchers thus are to be found largely only in the ranks of scientists themselves or, in academia, philosophers of science who tend to be marginalized within science. Among the canonical but questionable doctrines of modern science are the ideas that brain cells give rise to thought, the idea that genes are read-only and cannot be changed by proteins, the idea that Earth has an iron-nickel inner core, the idea that evolution proceeds by the gradual accumulation of random beneficial mutations, and the idea that life defies a universal tendency toward disorder. Of course not all scientists believe (or even know of) all these ideas, and many people contest some of them in accord with their own understanding, personal beliefs, or, despite the rise of science, religious faith. Then there are also marginalized "new age" and political beliefs, for example, alternative therapies for cancer, proposals for a non-HIV etiology of AIDS, and "conspiratorial" questions raised about governmental versions of terrorist events. In all these cases the stakes for the spirit of science as independent critical thinking are raised.

TRANSFORMATIVE FAILURES AND MODELS

"Failure is more beautiful than success," poignantly wrote the novelist John Fante (1909–1983). And failure—a core process of learning, because it provides the negative feedback needed to adjust our ideas—*is* unfairly stigmatized. Learning itself is a matter of mistakes, as is, arguably, the evolutionary process, where new forms of living organization, successful variants, only exist and thrive against a background of less successful forms. So, too, from the profusion of neuronal and synaptic connections, much richer in the newborn infant than in the adult, unused pathways are weeded out in what the Nobel laureate Gerald Edelman has termed neural Darwinism. These unused connections that in essence atrophy also represent a kind of mistake-making algorithm, the neurophysiological substrate and context of perception.

The British economist Tim Harford, in his *Adapt: Why Success Always Starts with Failure,* has emphasized the social cost of the stigmatization of error, despite its status as the core context for progressive learning both personally and socially. Harford contrasts the relatively conservative ethos of funding practices at the NIH to the more mistake-welcoming institutional culture of the Howard Hughes Medical Institute, which makes more mistakes but is measurably more innovative in terms of patents and medical breakthroughs. The willingness to fund creative, quirky, eccentric, experimental, and maverick people who sometimes flout guidelines, color outside the lines, follow hunches, or make counterintuitive bets sometimes pays off big. But within institutional practices tending toward ossification and groupthink, such outliers are rarely nurtured; it may be impossible to differentiate genius from madness, and their flouting of convention can be considered irritating, inappropriate, or even threatening. The paleontologist Martin Brasier recounts how as a ship naturalist his relative contributions were so great that he was told to take it easy, the message being that he was making others look bad by comparison. In some institutional cultures, academics are warned not to evaluate prospective candidates too highly for similar reasons. (The great fictional exploration of this theme is Kurt Vonnegut's story "Harrison Bergeron," about a talented boy forced to fit in by wearing a helmet that banged his head, bringing his cognitive powers into line with the rest of the socius.) This tendency to reduce to the common denominator can be seen as a detriment to the production of variety, the "mistakes" on which learning and evolution thrives. I remember the former editor of the *Scientific American* book review column, the nuclear and astrophysicist Philip Morrison, quipping, "An expert is someone who makes all possible mistakes." Morrison, who worked on the U.S. top-secret wartime Manhattan Project to build an atom bomb, physically helping load Fat Man onto the plane that dropped it on Nagasaki, later regretted his participation in that history-changing event as a mistake.

Science of course is arguably the greatest example of cultural learning based on mistakes. Unlike philosophy and religion, it formally integrates mistake making into its methodological structure. Although cultural historians and those in the discipline known as science studies dispute the official account of the scientific method—to forward and test hypotheses against the facts, and change the hypotheses if they don't match the facts—as oversimplified or an idealized caricature, there is certainly some truth to it.

The Viennese philosopher Karl Popper is well-known for his description of the scientific method as being about falsification: A bona fide scientific theory must allow itself to be proven wrong; otherwise it is an untestable ideology and not real science. What is less well known is the biographic component of Popper's influential theory: As a youth he was a committed communist, and when his friends analyzed the death of a close friend and comrade according to Marxist theory, Popper realized Marxism wasn't science.

This close friend of Popper's group was killed in a political demonstration in 1919 after some of Popper's friends came to the aid of fellow communists attempting to escape from the Vienna police station. Although unarmed, several of the young socialists and communist workers were shot and killed. Popper felt horrified and guilty, and partly responsible. This was the same year that scientists in Britain had experimentally verified Albert Einstein's relativity theory by measuring the curvature of light from another star as it passed by our sun during a total eclipse. Marxism, too, was presented as science, but was it?

"Marxist theory," wrote Popper, "demands that the class struggle be intensified, in order to speed up the coming of socialism. Its thesis is that although the revolution may claim some victims, capitalism is claiming more victims than the whole socialist revolution. That was the Marxist theory—part of the so-called 'scientific socialism.' I now asked myself whether such a calculation could ever be supported by 'science.' The whole experience, and especially this question, produced in me a lifelong revulsion of feeling."[13] Unlike Einstein's experimentally verifiable physics, there was nothing in Marxist ideology, no fact, experiment, or event, that could persuade the young Marxists that they were wrong. Popper, only seventeen at the time, came to believe that their friend had died in vain, for an unprovable belief system, not a verifiable science.

This sounds good in theory, but in practice the cutoff point between science and pseudoscience or ideology is more difficult to determine. Thomas Kuhn, in *The Structure of Scientific Revolutions,* argues that the progress of science is more dramatic, and discontinuous. A single or even a few counterfactual observations will not tend to bring down a vigorous scientific theory in one fell swoop. Rather, what tends to happen, according to Kuhn's analysis, is that anomalies, perplexities, and inconsistencies gradually accumulate but are tolerated until the old theory collapses. Such *paradigm shifts* (a term that has entered the vernacular) occur not because of being falsified in a simple way but because of sociocultural and factual factors. Kuhn acknowledges his debt to Ludwig Fleck, author of *Genesis and Development of a Scientific Fact.* Fleck shows how what are considered objective scientific facts often depend on the prevailing "thought-style" of a like-minded scientific community. Thus, in medicine (Fleck's specialty), some symptoms attributed to syphilis in fact resulted from the toxic mercury used to treat it. Popper's falsification idea is good in theory, but in practice scientists tend to interpret anomalies in terms of prevailing theories, ignore discrepancies as errors, or invent ad hoc subtheories to account for conflicting evidence that is too plentiful to ignore. A valid version of the scientific reception of a new theory is that there are three phases. In the first, it is dismissed as flat-out wrong. In the second phase, it is considered correct but trivial. Finally, it is considered both true and important—but those who initially objected to it now claim they had known it all along. Unless they are dead—nature's way of shaking up the dogma Etch A Sketch?—in which case they no longer have to be convinced.

Distressing as it is to those of us who would like to distinguish cleanly between truth and nontruth, science and pseudoscience, Fleck's analysis of hierarchical human thought-collectives seems to be largely on point. Aware of the lability of the status of facts framed within human groups, he might have enjoyed the word *factish* put forth by the French philosopher of science Bruno Latour. As I borrowed it earlier to discuss the new fact(ishe)s of life, a factish is not a fact of nature that exists as if it were isolated from human thoughts and perceptions; rather, it is something that we believe is real and independent, but if we look at it more closely we can see it depends on a specific historical scientific community and its thought-styles. One of Stengers's examples is the neutrino, which has never been seen in practice but whose traces on measurement devices within a certain context of physical investigations have sufficed to confirm its existence as objectively real.

Epistemology is the branch of philosophy that covers how we know what we know, and it is itself of course buffeted by many questions. The late atmospheric chemist Robert Garrels once told me, "Nobody believes anybody else's models; they are just a convenient place to store our data." This is a surprisingly pithy and revealing formulation, one that reminds us of the epistemological status of models—not as truths, per se, but as reserves of ever more abundant data. Garrels's homespun philosophy of science also highlights the suspect ascendancy of models in modern science. Models are meant to represent reality in the sense of simulating aspects of it, ideally in predictable ways.

What models are not is epistemologically ambitious. In other words, they have, in a sense, lost contact with science's guiding ideal, which is to find the truth, as Bohm says, "whether we like or not." Instead, models aim to make reasonable, mathematically grounded, simulacra of the physical interactions of objects. What they've given up, perhaps following the lead of quantum mechanics' Copenhagen interpretation, which suggests it's useless to seek a visualizable reality beyond mathematics' representation of subatomic particle behavior, is to find out "how things really are."

EARTH-CHANGING VIEWS

The maverick nuclear physicist J. Marvin Herndon laments this descent into a world of scientific modeling, which, satisfied with reasonably coherent abstractions, has lost touch with science's original mandate to discover, rather than just represent, reality. And Herndon points out, adding a somber warning note to Garrels's insider's view of scientific epistemology, that the ascendancy of modeling has pernicious consequences. Nor is it just modeling. The entire character of science has been changed by governmental support, corporate connectedness, and groupthink–nurturing institutionalization. In a sense it has always been this way. Neither Niels Bohr, supported by the Carlsberg brewery, nor Einstein, working in a patent office, received government money to make their great discoveries.

Herndon's analysis is not just academic: If Galileo was sufficiently cagey and diplomatic to espouse his views without (except for a little house arrest) undue fuss and ultimate acceptance, Herndon has not (so far) been so "lucky": Herndon's carefully articulated, evidence-based revision of the commonly accepted view of the geophysical history of our planet and its inner composition has until now been ignored or rejected by his professorial peers. Although quite complex in detail—his work dovetails astronomy, geochemistry, geophysics, and nuclear physics—Herndon's basic idea is both radical and simple.

He believes that Earth began its existence in roughly its present distance from the Sun but as a gas giant the size of Jupiter. The T Tauri solar wind, associated with the Sun's thermonuclear ignition, early on blew off Earth's atmosphere. This left a rocky kernel still compressed from the weight of about 300-Earth masses of primordial gases. Earth then began to decompress, driven primarily by energy stored from that great compression, energy that eventually split the surface into continents and wrinkled the crust to form mountain ranges. Herndon dubs this process, which corrects and extends plate tectonics, whole-Earth decompression dynamics (WEDD).

Herndon's picture of our Earth and its origin is a radical departure from the one that has been promulgated since the late 1930s. In 1936 the Danish seismologist Inge Lehmann, tracking changes in the angle of propagation of earthquake shock waves, discovered that Earth's core was separated into an outer liquid and an inner solid region. In 1976 Herndon published a paper, communicated by his teacher, the Nobel Laureate Harold C. Urey, a world expert in the cosmic composition and distribution of chemical elements who was an adviser in Stanley Miller's famous origins-of-life experiments, in the *Proceedings of the Royal Society of London.* After publication Lehmann sent a complimentary letter to Herndon. She was impressed by the logical development of his decompression idea, supported by meteoritic evidence unknown at the time that the standard geochemical model of Earth's origin and composition became accepted.

In the decades since he received the letter from Lehmann, never known to suffer fools gladly, Herndon has been hard at work investigating and developing his views. Inside Earth, he argues, the molten iron core is not simply cooling and solidifying as taught in textbooks, but its components and energy production derive in part from radioactive elements incorporated during its origin. The inner core, Herndon argues, is nickel silicide, and at its center is a "georeactor"—a powerful subcore of naturally fissioning uranium surrounded by a subshell of nuclear waste. It is this georeactor, and not the crystallization of iron leading to convection, that produces our planet's geomagnetic field as well as much of the heat affecting our planet's surface. With a mass only one ten-millionth that of the fluid core, the georeactor is capable of quickly switching the polarity of Earth's magnetic field. Indeed, it is capable of doing it far more quickly than if we assume, as standard geochemistry does, that geomagnetic field production takes place in Earth's massive fluid core.[14]

Herndon's analysis that Earth's interior has a solid inner core that is not iron nickel metal but nickel silicide, a nickel-silicon alloy, matches, unlike the standard geophysics model, the aggregate accepted density for our planet. Moreover, such a composition is strongly supported by so-called enstatite chondrites, meteorites mostly ignored after the standard model of Earth's composition and core became stratified, as it were, into geophysical thinking. Enstatite chondrites are distinct from the more common ("ordinary chondrites") kind formed under more oxidized conditions on which the standard model of Earth's origin and composition are based. They are more chemically reduced, closer to what we might expect of fragments left over from the relatively hydrogen-rich early solar system. Less available oxygen at Earth's formation meant that certain "oxygen-loving" elements unexpectedly occurred in the core; calcium and magnesium would have combined with sulfur and floated upward to form the thin layers detected between the core and mantle; and uranium, most likely combined with sulfur, would have sunk to the center. In Herndon's view silicon combined with nickel and sank to form an inner core of nickel silicide (not iron) and uranium concentrated at the center of Earth functioning as a nuclear reactor, while its churning layer of nuclear waste produced and continues to produce our planet's geomagnetic field.[15]

Today Herndon adduces multiple lines of evidence to support his claims, including similarity of Earth-parts with corresponding parts of oxygen-starved enstatite–chondrite meteorites and similar thermodynamic condensation considerations; similarity of calculated georeactor-produced helium with helium observed coming from deep within Earth; historic geomagnetic field reversals on a time scale of weeks to years; simultaneous historical pulses of lava production on opposite sides of Earth in Iceland and Hawai'i; and, more recently, deep-Earth geoneutrino measurements.[16]

Herndon's comprehensive whole-Earth view extends beyond core composition to nearly all current geophysical processes. And it is buttressed by new discoveries from the international space program, which has detected Jupiter-sized exoplanets at distances from their stars similar to that of our Earth from the Sun.

Is it true? Did our little rock begin with something like planetary delusions of grandeur? If so, what should we call it, this lifeless monster, this fat paleo-planet whose gases bore down and compressed the rocky inner part of Earth to some 64 percent of its present diameter, before T Tauri eruptions swept those gases away, not only from our planet but from Mercury, Venus, and Mars as well, and set the stage for little neo-Earth as we now know her?

Perhaps we should call it Jaea, this big old globe that none of us had expected, naming this mother after a combination of Jupiter and Gaea, the massive planet and the old Greek word for Earth. Or Jupaea maybe, or Juvea (which has a nice connotation of youth in addition to merging Jove and Gaea), or Megaea, or more simply Mega or Colosso, this proposed super-Earth (but this last term has already been reserved for extrasolar planets slightly bigger than Earth but with masses far below gas giants).

And it is not trivial, this matter of the name in the reception of an idea. Bumper stickers say "Reunite Gondwanaland," but the supposed supercontinent floating alone in the world ocean never existed if Herndon's ideas (which fascinated my mother, of course, who encouraged Herndon to better explain them) hold true.

I think I like Jovea.

TO UNDERSTAND WHY, despite more than thirty-five peer-reviewed professional publications, Herndon continues to suffer rejection without valid criticism for which he would be grateful, Herndon delved into the history and philosophy of science. Although he was not a sociologist, anthropologist, or philosopher, he strove to understand why his colleagues refused to grant a fair hearing to what he considered his well-documented alternative and compelling concept of Earth's formation.

Herndon blames, in part, the peer review system, instituted in 1951 at the end of World War II by the U.S. National Science Foundation. It removes personal accountability. In a shroud of anonymity, many reviewers nix projects that threaten work in a "thought-style" that differs from theirs. Does not admission of any error in a large shared theory ultimately threaten current authorities and, crucially, therefore menace funding?

"Prior to World War II," writes Herndon, "there was little government financial support for science. Nevertheless, the 20th century opened and seemed to offer the promise of an unparalleled age of enlightenment and reason. Fertile imaginations put forth ideas that challenged prevailing views. New understanding began to emerge, sometimes precise, sometimes flawed, but tending toward truth they inspired more new ideas, continued debate and further imaginative creativity. Enthusiasm and excitement ensued in the general public and kindled the imaginations of the young. Although money for science at the time was in short supply, scientists maintained certain self-discipline. A graduate student who worked on his Ph.D. dissertation was expected to make a new discovery to earn that degree, even if it meant beginning again because another made the discovery first. Self-discipline was also in the scientific publication system. A new, unpublished scientist obtained criticism and endorsement of published scientists before submission of manuscripts. The repressive popularity-contest system of 'peer review' had not yet been born."[17] The facelessness, pseudonymity, and anonymity of the Internet are similar to peer review. Science thrives under a regime of openness, fair and accountable judgment, not corporate patents, self-interested censorship, and informational control.

How do we safeguard the scientific spirit that has been so prodigious in our understanding of ourselves and our material environment? That mode of inquiry, that mix of close observation, childlike curiosity, and critical thinking that has enriched our inner lives and asymptotically approached the elusive philosophical ur-goal of self-knowledge? Not only is this unique modern heritage, arguably culturally blind at its core but ethnically inflected—this scientific spirit that is a universal birthright of all human beings and which is ultimately bigger than humanity—not only is it fighting for its life, pushed and pulled, captured and abused, its knowledges horded and copyrighted and patented against its antisecrecy divulgatory essence; not only has it become the plaything of corporations and publicity stunts and media and manipulators and lawyers with no intrinsic allegiance to the truth over and above their clients' cases and the accumulation of an all-too-human wealth; but by turns well-meaning, playful, and fashionable impulses within academia have also played into the hands of those who would willfully obstruct the search for open and universal knowledge. Without impugning their contributions, it is clear that the movements variously and collectively known as post-structuralism, deconstruction, and Continental philosophy have put scientific truth, to use one of their own terms, "under erasure." Jacques Derrida's able and in many ways useful critique of a "transcendental signified" is part of a climate where radically distinct perspectives are granted a sort of politically correct intellectual equality. While in some cases this may be a valued corrective to ideology posing as truth, it also provides a license for machinators and the benighted to claim ill-gotten gains. Add to this the difficulty of distinguishing between specialized discourse, the technical jargon of scientific specialties, and linguistic protectorates with in-crowd shibboleths, and we plunge into a dizzying abyss of unwatched watchers and incomprehensible experts.

The tension between the scientific attitude and the scientific reality is manifest in science's status as a collective enterprise dependent for the most part on public funds. Corporate and state backing of the search for knowledge is problematic in part because collective human organizations, according to evolutionary logic itself, are in the business of self-perpetuation. Assuming even a multivalent, situated definition of truth, for example, a truth of relativity if not an absolute or absolutely relative truth, there is still no reason various levels of obfuscation and deceit can not only prevail but be incorporated into structures of group perception and survival. Epistemologically, truncation, editing, and abstraction are necessary at even the most basic levels of perception. While the metaphor depends on contemporary technology, consciousness itself can be parsed as an "operating system" that hides the working innards of neurochemistry equivalent to wiring or hardware of a meat machine. The brain as interface, altered by electrochemistry, in this view would not command an absolute truth but a workable one. Here we encroach on the epistemological compromises elucidated by pragmatists in philosophy, to wit, that workability makes its own truth. If we cling to a classical definition of truth as an abstract ideal—certain, universal, necessary, and true in its mathematical ideal—we run afoul of the evolutionary process, which does not necessarily move in the same direction. This is especially true when one considers the holarchical or nested nature of groups consisting of perceiving individuals who or which may themselves be dispensable in terms of a collective survival enterprise. This is the aporia of evolutionary epistemology: that knowledge of truth in no way ensures survival. In fact, it can be directly inimical to it. Such is the background behind the legend of achieving complete knowledge only upon death, of stories where final secrets are revealed taking those to whom they are whispered with them. It is perhaps ironic that the emphasis on selfish genes in neo-Darwinism, and the virulence reserved for the supposedly mathematically untenable theory of group selection, are maintained by coteries of group-acting beings, that is, by old boy networks of the academic kind, intellectual cabals of browbeaters quick on the rhetorical draw and not above using straw men, ad hominem arguments, and other tricks of emotional persuasion in contradistinction to rational analysis.

Protecting the spirit of science and safeguarding its path are not trivial in our tribal species in the current politico-corporate and academic environments. The scientific attitude comes out of a deeper philosophical stance of curiosity and critical thinking, a stance that must face down dogma even at the risk of social disapprobation. In a media age, in a public relations state where science can and has been corrupted by corporate bottom lines and government-sponsored agendas, the philosophical heart of science—thinking things through for yourself and seeking the truth whether or not we like what we find—is increasingly in peril.

Groupthink intrinsic to tribal survival in combination with the displacement of the knowledge priesthood from the clergy to (courtesy of Bruno and company) lay and scientific authorities can make for a volatile—and distinctly scientistic, that is, antiscientific—mix. In the binary oppositions of the dichotomizing, such authority may seem to be the opposite of pseudoscience, but in fact it is equally the opposite of science, all the more dangerous for the institutional power of authority it wields.

> No snowflake in a snowstorm ever feels responsible.
>
> —*Voltaire*

NAMES ARE NOT THINGS, but they come mighty close.

"What's in a name?" asks the star-crossed Juliet, feminine force of the play that bears her name. "Art thou not Romeo, and a Montague?"

"Neither, fair maid, if either thee dislike."

Juliet is right to worry about her lover's name, and the contentious family history it signifies cannot just be wished away. There is the love of Romeo and Juliet, and there is what society makes of that love, and however much society betrays the essence of that love with its fatal expectations, those expectations, reinforced by would-be arbitrary names, cast a real pall over the young lovers' romance.

The name, if not quite a magic word making something that was not yet there jump into existence, creates a kind of field, organizing our expectations and creating real-world effects, for better or worse, according to a logic of self-fulfilling prophecy. Romeo should be against Juliet because she is a Capulet; Gondwana and Pangaea, with their proper names, must refer to real supercontinents that once existed.

HIV is another, highly problematic, example. Leave aside any medical protocols that were broken (and that HIV's founder, Robert Gallo, was censured; and that the discovery of the putative virus was never made according to the Koch postulates hitherto considered crucial in virology; and that there are apparently no electron micrographs of the virus; and that tests identify antibodies that can be to a variety of proteins; and that Peter Duesberg, whose career was effectively ruined for questioning the HIV-AIDS connection, offered to inject himself with HIV to prove his deep doubt; and that Luc Montagnier, one of the Nobel Prize winners for the discovery of HIV, has become a target for derision because he does not espouse the HIV-AIDS party line),[18] and just for the moment consider the acronym: Human Immunodeficiency Virus. At once subtly and blindingly obviously, the name contains the answer to the question we are supposed to be still asking. Like Molière's "dormitive principle" that was a sufficient explanation for why we sleep, but in a tragic rather than comic register, the name simply restates the desired answer while foreclosing the question. Now no one in their right mind wants to wade into the vat of retroviral sludge and priestly accusations of denialism and conspiracy theory and weak minds that is this politically loaded issue. That's why I'm not going to get into this here. I have smaller fish to fry.

IN HERNDON'S VIEW, our planet's decompression comes about as a natural rebound from the crushing compression caused by the great overburdening weight of the gas giant stage. Earth's surface area must increase to accommodate expanded planetary volume, and it does so by "rifting," forming surface-splitting cracks. Cracks with underlying heat sources produce lava that forms the midocean ridges and paves the ocean basins before ultimately falling into and in-filling cold decompression cracks observed as oceanic trenches in a process that explains ocean floor geology without assuming mantle convection. Moreover, georeactor-produced heat, channeled to the surface, powers hot spots such as Iceland and Hawai'i and aids in continent fragmentation, such as presently occurs at Afar in the East African Rift System.

This last Herndonian tenet is especially contentious (for those who even know about it), because plate tectonics was itself a hard-won, revolutionary geological idea, itself one of the poster children of scientific revolutions. Bold scientific thinking seems to be somewhat addictive, as if removing the chains of institutional decorum allowed one to run free through the Elysian fields. One book by Herndon, emboldened by the prize-winning journalist Guy Gugliotta's accolade that he is a "Maverick Geophysicist,"[19] is even named *Maverick's Earth and Universe.* (If you are interested in looking at other mavericks and their claims, and attempts to judge them outside received opinion, in a contemplative noncommercial setting, I recommend the website Science Guardian: Paradigms and Power in Science and Society.)[20]

Obviously the tenability of Herndon's connected suite of geological ideas won't be settled here, but a couple of aspects of them are highly relevant to the whole question of scientific epistemology, of science-in-the-making and how the layperson, or indeed specialized scientists themselves, are supposed to adjudicate new possibilities of "how things really are" in the modern environment of institutionalized, governmentally supported, and fundamentally conservative science.

The inertia of model making as opposed to fresh consideration of conflicting evidence is itself one of Herndon's pet peeves. Although the division is no more ironclad than that between science and pseudoscience—Einstein initially dismissed quantum mechanics for its "spooky action at a distance" (a criticism that, when you think about it, might also be applied to Newtonian gravity!)—Herndon is keen to differentiate between model making and real science. And one signature of model making, with its emphasis on mathematical simulations, is the tendency for model makers to absorb would-be counterfactual evidence into their model by making ad hoc hypotheses to accommodate them. One of the more famous historical examples is from Ptolemaic astronomy. Observations of the outer planets from Mars to Neptune with Earth as the assumed solar system's center must account for these planets appearing to move backward or "retrograde" (still familiar talk in astrology) for two to six months at a time.

This artifact of a "mistaken" or at least not as elegant frame of reference is reminiscent of the wagon or car wheels appearing to spin backward in films, because of a discrete number of frames per second in which the wheels fall short of a complete revolution before being caught on film. In Ptolemaic astronomy the backward movements were handled by the ad hoc hypothesis of epicycles—perfect circles that moved around off-centered points (equants) of a circle (deferent) around Earth, with each planet having a separate set of parameters and the center of the orbit never exactly Earth itself. A good example of a model that reproduced observations but wasn't really true, the Ptolemaic epicycles (though Copernicus also used them, because it wasn't yet realized that the planetary orbits were elliptical) might be compared with a description of your car going backward or "retrograding" when another car passes you on the way to work: descriptively accurate with regard to your chosen frame of reference, but not exactly true.

Strikingly similar, one might argue, is Herndon's still-heretical view, based on Earth's enstatite–chondrite composition, addressing the density anomaly in the standard model, accounting for the tendency of the geomagnetic field's polarity to flip (much easier if generated by a georeactor), which was published and developed *before* the discovery of Jupiter-sized extrasolar planets close to their suns. The discovery of such giant near-to-sun planets has caused some scrambling amid the well-funded planetary model makers. Herndon's view is comparable in its intellectual housecleaning way to Copernicus's repositioning of the Sun to be at the center.

The relevance for modeling here is that Herndon's "heretical" but more elegantly integrative view does not require ad hoc notions such as recently proposed "planetary migration" (of big outer planets to close-to-sun orbits) to account for the new evidence of Jovian-size extrasolar planets traveling in the range of one astronomical unit, or about the distance of Earth from our Sun, around their stars. Indeed, recently detected antineutrinos with a nuclear reactor spectrum coming from inside Earth suggest that as much as 26 percent of deep-Earth heat from uranium and thorium is produced by the georeactor (15 percent, according to Italian data).[21]

Herndon differs from the geophysicist's tradition in that he rejects the assumption that our changing Earth has maintained a constant diameter. Continental drift and plate tectonics *assume* that Earth's diameter has remained constant at today's value through time. Herndon argues for a "continental cracking" of the originally continuous lithosphere as Earth decompresses. Hot cracks are forced to widen as the lithosphere breaks and spreads. Seafloor lava, pumice, and basalt spew out. Cold cracks, recognizable to geophysicists as their "subduction zones," open to become the ultimate repositories for the basalt and the sediments that ride in on it.

Here, derived from Earth's early origin as a Jupiter-like gas giant, is a new geoscience paradigm that explains myriad observations attributed to plate tectonics. Herndon's explanation rejects "mantle-heat convection theory," for which he claims there is very little evidence. Proposed by Alfred Wegener, John Tuzo-Wilson, Maurice Ewing, Frederick Vine, Drummond Matthews, William R. Dickinson, and others, the grand "continental drift–plate tectonics" paradigm plate-tectonics vision is itself a poster child for interdisciplinary scientific revolutions. Will future scientists look on in dismay at the rigid institutional structures Butler saw emerging, roadblocks that prevented Herndon's connected suite of professional geophysical concepts of our "indivisible Earth" from being considered?

Herndon is not the only victim of rigid thought-styles. Another may be the thirty-year rejection of the "heretical" science of the marine zoologist Donald I. Williamson, working out of the Port Erin Marine Station on the Isle of Man. The English historian of biology and medicine Frank P. Ryan tells Williamson's story of "saltatory evolution" in his recent works on metamorphosis, and the work is also documented in a 2011 film *Hopeful Monsters* by Robert Sternberg of Imperial College London.[22] Animal larvae may evolve by hybridization: fertile crosses between members of different animal phyla may account for extraordinary metamorphoses in the sea. Despite intriguing evidence for these claims,[23] there is extreme reticence among professional evolutionary biologists to even print Williamson's ideas. In detailing evidence for evolution by hybridization, Williamson explicitly disputes Darwin's central tenet that "descent with modification" (as he termed evolution) occurs slowly by gradual accumulation of naturally selected variations.

Darwin's view was made in contrast to Christian views of special creation. Darwin, in *The Origin of Species,* acknowledges Aristotle's knowledge of natural selection but dismisses the philosopher and early biologist's views. Ironically, however, Aristotle dismissed natural selection because he in turn associated it with Empedocles's pre-Socratic myth that organs once roamed Earth on their own, occasionally joining up and fusing as they were naturally selected into new forms. Aristotle seems to have thrown "the baby out with the bathwater," and Darwin to have rejected rapid evolutionary change because it smacked of special creation. In both cases we can see these brilliant minds overreacting to protect against the excesses of their ancestors: in Aristotle's case, against mythic accounts of chimeric unions among men, animals, and gods that violated observations of nature; and in Darwin's case against the notion of simultaneous creation of species, the monotheistic explanation. Not even the greatest scientists, aware of the need for evidence, are immune from a tendency toward dogma. Ideology inhibits inquiry. New myths are created as people overshoot in their zeal to refute old myths. Everybody needs something to believe in, even nonbelievers.

Gazing into the spiral of the history of science is a fascinating spectacle: not only do "truths" become "myths," but, as the spiral of evidence unwinds, some of what formerly was dismissed as myth, with new observations and new methods by new investigators under changed social conditions, sometimes is resuscitated as scientifically valid, with qualification, after all. Science nurtures endless curiosity and further investigation, exploration, and description. Really new scientific truths have come from kooks who need criticism and suffer failures. Beyond the battles of cranks and dogmatists dug deep in their holes are the alliances of skeptics and the curious, able, if evidence warrants, to change their minds. Reexamination, reinterpretation, reevaluation, and reinvestigation are intrinsic. The progress of knowledge through a combination of critical inquiry and open-mindedness is science itself.

CODA

After my mother died, her longtime companion, the highly respected Spanish microbiologist Ricardo Guerrero, when I asked him about the character of her rebellious views, mentioned that, though she's an intellectual heroine in Spain, there were three things about her he refused to discuss there: (1) 9/11, (2) the *PNAS* affair (in which she, availing herself of her privilege as a member of the National Academy of Sciences, had Williamson's paper on fertile cross-species unions published in the *Proceedings of the National Academy of Sciences*,[24] and (3) AIDS (where she questioned the canonical HIV-AIDS connection). When I asked him what he thought of her advocacy of Gaia theory, however, he quickly answered that it was one of the greatest scientific theories of the twentieth century. This contrast between Guerrero and the neo-Darwinists (whom Gould called "Darwinian fundamentalists") exemplifies the existence of the kook–critic continuum of my subtitle: the very progress of science depends on tireless questioning of received opinions, especially when they are both supported by evidence and threaten entrenched assumptions, money flows, and careers. Truth, or its asymptotic representatives, may be stymied but in the long run "will out," as Shakespeare put it.

CHAPTER 12

METAMETAZOA

LIKE A GRAY GEODE CRACKED OPEN to reveal coruscating crystals of amethyst, the history of science sometimes surprises. Empedocles imagined an ancient world of organs mating and merging with one another to create bizarre half-hewn beasts, the most favorable matches surviving. Aristotle, schooled in Platonic typology and sick of unlikely stories of cross-species mating, metamorphosed mortals, and shape-shifting gods, rejected Empedocles out of hand.

But the chronological vortex of knowledge's wayward march turns on itself like a DNA molecule: Now we know that Aristotle, first biologist though he may be, was wrong on both counts. Empedocles's intuition of natural selection and symbiosis was on the mark. Nothing as ghoulish as crawling pancreas and self-pumping hearts getting it on in the primordial mud, but today's textbooks teach that our organelles, parts of the cell outside the nucleus, once swam as beings on their own. These are the mitochondria, and they give you the genetic intracellular infrastructure to breathe oxygen, a toxic waste gas first released in massive quantities into the atmosphere by cyanobacteria, which came up with the clever idea of using water to grab their hydrogen atoms. The archaea victimized by bacterial air pollution were saved by bacterial infection. We now celebrate their double victimhood with every breath we take, as their infection evolved into mitochondria at the heart of all animal metabolism and energy use. In retrospect, Empedocles was right. Although there's no evidence of his "man-faced ox-progeny" that "perished and continue to perish,"[1] there are, we could say, "bacteria-bodied archaea-progeny" that "survived and continue to survive." You are one of them!

A similar story could be told of the scientific disproof of spontaneous generation. Lazzaro Spallanzani and Louis Pasteur with curved glass flasks protecting inoculation proved that life doesn't arise from nonlife—meat doesn't beget maggots, mice don't defrag from rags. Yet, flash forward once more, to the origins-of-life experiments showing that amino acids naturally form when ammonia and other hydrogen-rich compounds are exposed to an energy source, and the possibility arises that life *can* occur from nonlife and not just be seeded by spores from the air (or space; see chapter 4). As in the T. S. Eliot poem, we return to a different floor of the expanding vortex, the Escherian staircase of science. Another example is the great Sphinx at Giza, whose chimeric mancat body is made of stone, yellowish calcium carbonate studded with nummulites, "coin stones" that Herodotus thought were fossilized lentils but are actually the remains of foraminifera. Live forams fill the oceans, their tiny spiked carbonate bodies usually no bigger than a pinhead but sometimes growing to inches without giving up their status as single cells. Cut transversely, they show spirals. And many of their species, whose variety tracks the hidden treasure of fossil fuels, are symbiotic: Again the baroque vortex turns, the mythological blend that is the Sphinx is a creature of the imagination, is made up, but its real body is made up of the fossil remains of real chimeras, as are the Great Pyramids themselves, 40 percent of whose yellow limestone consists of fossil forams, many of them symbiotic with specific species of diatoms and dinomastigote algae that floated in the Tethys Sea, during the Eocene, more than thirty million years ago.

The oldest known sphinxes are from Anatolia, Turkey, and are over nine thousand years old. The oldest fossils of possibly chimeric beings, such as acritarchs, are over three billion years old. More recently, in attempting an untested sleight with a piece of Dominican amber, I dropped it on the floor of my mother's lab at the Morrill Science Center at the University of Massachusetts. This missed trick with the fossilized tree sap on loan from David Grimaldi of the American Museum of Natural History paved the way for thin-section electron microscopy.

The cracked amber revealed a twenty-million-year-old termite whose hindgut was full of petrified swimmers--spirochetes, cellulose-digesting protists, and the spore-forming filamentous bacterium, *Arthromitus*—equivalent to *Bacillus cereus,* identical except for two or three plasmids, which are DNA coiled into tight rings, to *Bacillus anthracis,* the causative agent of anthrax.[2]

Life's tenure on this planet has been so long, and its grip so tight, that many things that seem to be singular organisms reveal themselves—like the Sphinx or that tawny piece of Miocene butterscotch cracked open like Humpty Dumpty to reveal the termite *Mastotermes electrodominicus,* now known only as a living fossil from northern Australia, or the fossil itself, containing a miniature Pompeii in its swollen hindgut—to be constructed of other life-forms. As the helix of history turns, we descend deeper along this escalator into life's ancient forms, its ghostly archive and living tombs. It slipped out of my hand, but that magical piece of amber made the cover of *Science News,* one of science's leading magazines.[3]

HOW WILL THE BODY AND ITS LIFE come to have been construed by the future of biology? And more important, how will these be construed by a mythopoiesis and popular mythology whose social birthright now, through the midwife of contemporary biology, may create the "facts" from which a common future understanding will come? Transformations of classical models are already under way in contemporary biology, and here I look at three of them: Gaia theory (geophysiology or Earth system science), symbiotic evolution (symbiogenetics), and bacterial omnisexuality (hypersexuality).

It is necessary, first of all, to distinguish the tenor of a "new biology," whose theoretical sources are Gaia, symbiosis, and gene-trading bacteria, from the tenor of the more traditional biology for which the paradigm of individuality is the animal body. Modern biology, informed by cellular ultrastructure through electron microscopy and detailed knowledge of gene sequences, has supplemented or even negated the long-standing division between plant and animal kingdoms.[4] Although vying for acceptance and mutually inconsistent, the two most favored current phylogenies split life into either three domains or five kingdoms. The five-kingdom classification system still reserves a place for the kingdoms Plantae and Animalia (both subsumed within the superkingdom Eukarya).

Carl R. Woese's three-trunked tree of life, based on typical sequences of RNA in the ribosomes of cells, contains no separate kingdoms for plants or animals, for it lumps both within the eucarya (organisms composed of cells with nuclei), reserving two separate taxa (archaea, which Woese used to call archaebacteria, and bacteria, formerly eubacteria) for the rest of life. Molecular biology and microbiology have not only confirmed Charles Darwin's paradigm-shifting argument that we are animals but have also provided evidence that the most fundamental fence in life lies not between plants and animals but between eukaryotes—cells with nuclei, mitochondria (and, in the case of algae and plants, plastids)—and prokaryotes, also known as monerans or bacteria. *Homo sapiens* clings to its crown as the walls of its kingdom come crumbling down. Moreover, each eukaryotic "animal" cell is, in fact, an uncanny assembly, the evolutionary merger of distinct prokaryotic metabolisms. Strictly speaking, there is no such thing as a one-celled plant or animal.

Easily recognizable life-forms appear only at the middle range. If we step back from, or come closer to, the living canvas, organisms blend into a pointillist landscape in which each dot of paint is also alive. In short, all previous biology has been grossly zoocentric.

Although psychoanalysis and phenomenology and their popular offshoots have disturbed a monolithic conception of mind, a monolithic notion of "the" body remains largely intact. In classical medicine, the body is considered a type of unity. Cancer, paradigmatically, but other diseases as well, are discussed with the rhetoric of war: the body is "attacked" and "invaded"; it puts up "defenses" and "fights back." This medical model of the body-as-unity-to-be-preserved, though, of the body proper, is besieged by the new biology.

A radical rerendering of the body is under way in accordance with three models from the new biology, namely, symbiosis, Gaia, and prokaryotic sex. This reformulation augurs a breakdown of the medically proper animal body, which is simultaneously driven in at least two new directions, one post-structural and the other medieval-microcosmic in terms of extended selves within a living environment. Gaia refers to the biosphere understood not as environmental home but as body, as physiological process. Prokaryotic sex, or bacterial omnisexuality, refers to the fluid genetic transfers, by definition sexual, among continuously reproducing bacteria.

The consonance with certain post-structuralisms occurs in that the new biology parts company with the unitary self assumed in the zoocentric model. The expression "medieval-microcosmic" is inadequate but suggests the possibility of correspondences among prokaryotic, eukaryotic, zoological, and geophysiological (Gaian) levels. It now appears that a type of individuality has appeared at each of these levels. Both spatially and temporally more inclusive, Gaia and animals dwell within a holonomic continuum, superordinating the smaller beings of which they are made.

THE BODY AS CHIMERA

The body as seen by the new biology is chimerical. Instead of the tripartite division of that mythical creature of antiquity, the chimera, into lion, goat, and snake, the animal cell is seen to be a hybrid of bacterial species—although the word *species,* as seen below, is not that apt when applied to bacteria. Like that many-headed beast, the microbeast of the animal cell combines into one entity bacteria that were originally freely living, self-sufficient, and metabolically distinct. Mitochondria populate and energize virtually all eukaryotic cells. These specialized cell parts respire; they take up oxygen and produce carbon dioxide, making ATP (adenosine triphosphate), a kind of molecular capacitor storing energy within cells. It is now widely accepted among biologists that these tiny intracellular power stations were once autonomous respiring bacteria. Eukaryotic cells evolved over a billion years ago, probably when respirers entered and did not kill but *were incorporated by* larger anaerobic archaea. The archaea include sulfur-breathing, acid-resistant, and heat-tolerant extremophiles—an impressive range of resistances that may reflect an ancient ability to survive hot temperatures and meteoritic bombardments of the early Earth. Over time, the two distinct metabolisms merged, and the new incorporated cells produced more and hardier cells than either line of their unincorporated relatives.

Some intriguing signs recall the ancient free lives of mitochondria. Although they lie outside the cell's nucleus, they have their own genetic apparatus, including their own DNA, messenger RNA, transfer RNA, and ribosomes enclosed in mitochondrial membranes. Unlike the DNA of the nucleus, but like bacterial DNA, mitochondrial DNA is not coated by histone protein. Mitochondria assemble proteins on ribosomes very similar to the ribosomes of bacteria. Both mitochondrial ribosomes and those of respiring bacteria tend to be sensitive to the same antibiotics, such as streptomycin. Perhaps most telling, mitochondria reproduce on their own timetable and in their own way, forgoing the complex mitosis of the nucleus for a simple bacterium-like division. They engage in the nonsystematic genetic transfer that characterizes bacterial sex. All in all, they behave like prokaryotic captives.

As early as 1893, the German biologist A. F. W. Schimper proposed that the photosynthetic parts of plant cells came from cyanobacteria (often still called blue-green algae, but the term is a misnomer, since they have no nuclei in their cells). The French biologist Paul Portier believed by 1918 that mitochondria are the descendants of bacteria that had become lodged within the cells of animals and plants. In the first quarter of this century, the American anatomist Ivan Wallin and the Russian scholar-biologist Konstantin S. Mereschovsky had independently come to the same conclusion. In 1910 Mereschovsky, who taught at the University of Kazan, published an essentially contemporary view of the origin of eukaryotic cells from various kinds of bacteria.[5]

Experiments at isolating the putative bacterial partners, however, have always failed; the evidence for cooperation rather than parasitism was overlooked and dismissed as "sentimentalism." Herbert Spencer equated the necessary evils of competition with an eminently desirable social progress, and Thomas Huxley referred to the animal world as a "gladiator's show"; Pyotr Kropotkin wrote *Mutual Aid,* and others implicitly linked evolutionary ideas of symbiosis to labor unions, mutualistic societies, and socialist ideas.

The animal cell, as well as the cells of plants, fungi, and protoctists (a miscellaneous eukaryotic kingdom composed mainly of algae and what were once called protozoa),[6] combines oxygen-using mitochondria and a larger host cell. Despite its suggestive appearance, the nucleus was probably never autonomous but, rather, the result of interactions among members of cellular communities that evolved into cells. The same cannot be said of the chloroplasts of plants and the plastids of photosynthetic protoctists such as algae. The grass-green photosynthetic organelles of all plants may be the descendants of a single, wildly successful bacterium, now shackled, albeit gently, in its cytoplasmic prison.

A body of behavioral evidence similar to that for mitochondria supports a cyanobacterial origin for the pigmented bodies within algae and plants. The ancestors of all plants were probably cells with mitochondria that ate, but never digested, their live vegetarian dinner. The undigested photosynthetic organisms grew inside their hosts, offering a steady diet of metabolites in return for protective cover and continued life.

The red plastids of seaweeds also probably come from autonomous bacteria. If one compares the sequence of nucleotide bases in the ribosomal RNA of red plastids in the seaweed *Porphryridium* with that of RNA in the seaweed's own cytoplasm, the resemblance is less than 15 percent. Making the same comparison with the ribosomal RNA of the cyanobacterium *Synechoccus* and the plastid of the swimming green protist *Euglena* yields similarities of 42 and 33 percent, respectively. Indeed, there have always been behavioral clues to the xenic origins of the eukaryotic cell.

But the striking likeness between mitochondria and respiring bacteria, on the one hand, and between plastids and photosynthetic bacteria, on the other, has "proved" beyond a reasonable scientific doubt that all cells with nuclei, from a unicellular amoeba to a multibillion-cell anaconda, come from more or less orgiastic encounters (eating, infecting, engulfing, feeding on, having sex with, and so on) among quite different types of bacteria.[7] It is now generally agreed, not to mention taught in textbooks, that both mitochondria and chloroplasts derive from bacteria.

Like the chimera, the plant cell recombines three distinct and once-separate entities: protective anaerobic host cell, internally multiplying photosynthetic bacterium, and respiring bacteria. The evolution of these last was sparked by the accumulation of highly combustible originally poisonous free oxygen within the atmosphere of the early Earth. Only anaerobic bacteria dwelled on Earth at this time. The lack of free atmospheric oxygen is attested to, some two billion years ago, by the replacement of banded iron formations (presumed to be the fossil remains of communities of photosynthesizing bacteria) with heaps of rust (representing the accumulation of atmospheric oxygen to near-present concentrations).

The buildup of atmospheric oxygen was itself a cyanobacterial phenomenon, since the metabolic waste product of the mutation, which allowed photosynthetic bacteria to use water as a source of hydrogen, was none other than gaseous oxygen, at first an extremely hazardous waste. Those bacteria that did not evolve to tolerate or use oxygen, or team up with cells that did, died. Although the accumulation within Earth's atmosphere of oxygen was initially catastrophic, it was also an energetic catalyst for organisms such as the oxygen-respiring ancestors of mitochondria, which employed the new abundance of the highly reactive gas to produce intracellular energy reserves at many times the rate of their fermenting bacterial predecessors.

Modern fermenting bacteria include anaerobes that, like trolls or elves in a cosmic fairy tale, protect themselves from the hazards of surface oxygen by dwelling underground. Vestiges of the anaerobic environment of the early Earth survive as shoreline stromatolites (rounded bacteria-built stones) and their softer relatives, microbial mats.[8] Human beings belong to the army of mutants that evolved in the aftermath of the oxygen infusion, the greatest pollution crisis Earth has ever known.

There is less evidence for another player in the symbiotic game whose evolutionary permutations have provided us with all known species of animals and plants: the spirochete. The spirochete's presence has been postulated to persist in ghostly form in all cells possessing the undulating organelles perhaps best known as undulipodia.[9] Undulipodia, whose electron microscopic ultrastructure reveals a characteristic "9+2" tubular form, are common to the cilia of women's oviducts and the sperm tails of ginkgo trees (and much else besides). My mother believed these undulating organisms derive from a not-yet-discovered species of spirochete. It was a seductive idea: these wriggly beings, so pervasive in oxic and anoxic muds, so prolific and with such a terrific penchant to form partnerships—sometimes attaching as swimming appendages to larger cells—moved in, like a sperm to an egg, but with more distinct consequences. If the ancestors to mitochondria can do it, and slow green beings that don't need to be taken to dinner, then why not the speeding spirochetes, so prolific in numbers and species, so metabolically versatile and naturally infectious?

Ever-squirming spirochetes fed not only alone but at the periphery and even inside larger cells. Inside cells themselves inside termites they can be seen to synchronize their slippery selves, so prolific is their reproduction in undulating activity in a limited space. In Grand Central Station swarms, these versatile beings, my mother reasoned, would be afforded all manner of opportunities for microbial merger; one possibility, she argued, was that neurological abilities made use of co-opted spirochete infrastructure, gene, and protein complexes, as this most ancient of eukaryote symbioses was redeployed in the neural and sensory apparatus (e.g., the active, microtubular brain) of "higher organisms."

The edges of larger cells, sites of leakage supplying a constant flow of food, were such prime real estate that some spirochetes appear to have renounced their former freedom of movement in order to permanently attach. Later, serving as a means of locomotion and food acquisition to larger cells, the spirochetes would have become increasingly phantomlike as they evolved ever-more harmoniously into the chimerical eukaryotic system. Today, mitosis and the mitotic spindle may be like the smile of Charles Dodgson's Cheshire Cat: the faded remnants of a life-form that has all but vanished into symbiotic thin air. In a living environment, parts of the self can be gradually lost. Spirochete remnants, alive and well, haunt the phenome, lying at the deepest, most ancestral levels of our being. Finding evidence for such spirochetes, however, remains a problem.

The human body, too, is an architectonic compilation of millions of agencies of chimerical cells. Each cell in the hands typing this sentence comes from two, maybe three, kinds of bacteria. These cells themselves appear to represent the latter-day result, the fearful symmetry, of microbial communities so consolidated, so tightly organized and histologically orchestrated, that they have been selected together, one for all and all for one, as societies in the shape of organisms.

The wastes of microbial communities, analogous to our garbage dumps and landfills, have also been incorporated as organisms breed together and organismhood appears at spatially more inclusive levels. The mineral infrastructure of our bodies, for example, the calcium phosphate of bone, owes its existence to the necessity of eukaryotic cells to keep cytoplasmic calcium concentrations at levels around one in ten million. Because seawater concentrations of calcium are often four orders of magnitude higher than this, ocean-dwelling cells must exude calcium to avoid poisoning. In the full-fathomed sea did the bones of our ancestors and the shells of their eukaryote relatives evolve. Skeletons dramatize an ancient waste, whispering a ghostly testimony to the useful internalization of hazardous waste sites. This gives a good indication of how life within the general economy evolves to deactivate, sacrifice, and eventually incorporate the dangerous excesses that accrue from its solar growth.[10]

The implications of a new biology for that identity which arises epigenetically from a single protist-like fertilized egg cell and for the zoocentric, medically proper model are immense. The body can no longer be seen as single, unitary. It is multiple, even if orchestrated by vicissitudes and the need for harmony over evolutionary time. We are all multiple beings. Our chimerical nature is less obvious than *Mixotricha paradoxa,* a species of autonomous nucleated cell (i.e., a protist) that seems to be unicellular but, on closer inspection, is seen to consist of several different kinds of cells—among them internal oxygen-respiring bacteria and externally attached spirochetes that serve as oars. In addition *M. paradoxa* contains its "own" organelles, congenital undulipodia used as rudders that, along with the spirochetes, help propel it through a droplet of water. In the transformation from organism to an organelle, cell membranes meld and, ultimately, disappear, as organisms undergo intraorganismic genetic transfers (an example of bacterial omnisexuality) to become organelles.

The brain's neurons, rich in the tubulin proteins that form the walls of the cell fibers known as microtubules, may also be the highly modified remnants of intra- and extracellular spirochetal mobility systems.[11] Undulipodia all consist of microtubules built of tubulin proteins; some spirochetes contain tubulinlike proteins. If the body–brain is not single but the mixed result of multiple bacterial lineages, then health is less a matter of defending a unity than maintaining an ecology.

Whereas the zoocentric model causally ascribes diseases to organisms, the new biology recognizes that many putative disease agents—such as streptococcus bacteria and *Candida albicans* fungi—are normally present in the human biological system. As we move to a new model, the body becomes a sort of ornately elaborated mosaic of microbes in various states of symbiosis. The distinct presence of these microbes becomes noticeable only when festering and illness throw normal populations and metabolite turnover out of equilibrium. Drugs used to treat bacterial meningitis can kill the bacteria, but in doing so can upset the body's internal microbial ecology with the result that fungi, usually held in check, proliferate fatally in the cerebrospinal fluid.[12]

Moreover, disturbances of the body's normal microbial ecology do not, properly speaking, signal sickness so much as the emergence of difference and novelty. Like cataracts or the glaucomous decay of vision, which may lead an artist into new percepts of flowery fields, so the same *Treponema* spirochete associated with deterioration has been linked with remarkable mental feats and artistic productions—here, for example, by the writer Anthony Burgess:

> I became interested in syphilis when I worked for a time at a mental hospital full of GPI (General Paralysis of the Insane) cases. I discovered there was a correlation between the spirochete and mad talent. The tubercle also produces a lyrical drive. Keats had both. . . . I've been much influenced by the thesis of Mann's Doctor Faustus, but . . . some prices are too high to pay. There was one man who'd turned himself into a kind of Scriabin, another who could give you the day of the week for any date in history, another who wrote poems like Christopher Smart. Many patients were orators or grandiose liars. . . . Some of the tremendous skills that these patients show—these tremendous mad abilities—all stem out of the spirochete.[13]

The body is not one self but a fiction of a self built from a mass of interacting, supervening selves. A body's capacities are literally the result of what it incorporates; the self is not only corporal but corporate.

GAIA

Just as the technologies of microscopic apparatus have opened the way toward a view of the human organism as a massive microbial ecosystem, in which eukaryotic waste products such as calcium have been honed into the calcium phosphate architecture of the human skeleton, so too have telescopic technologies opened the way toward seeing Earth as a living entity, in which animals are not independent actors but organelle-like components within a functioning planetary physiology. It is the timidity of our view, the ecological cognate of both our existential separation and our lack of understanding, that leads us to underestimate the interconnectivity and complexity organisms cocreate with one another and the substances of the geosphere, atmosphere, and hydrosphere.

Gaia, in its vulgar but succinct version, claims that Earth is alive. Yet, with greater nuance and accuracy, it can be stated that Earth's largely biogenic surface, the biosphere, appears to regulate itself physiologically within the astronomic medium; it behaves *as* a body. Gaia need not be conscious to act in ways that appear to humans to evince consciousness—even a computer program with complex feedbacks can mimic intelligence, and biospheric biological feedbacks are arguably far richer, the living crucible in which human intelligence evolved. Moreover, Gaia, while alive, differs from an organism because it is a closed ecology, recycling its wastes completely. This basic brutal fact also belies the notion that the Gaia hypothesis represents a fine and happy balance of harmony-optimizing organisms. There is a sense in which Gaia optimizes energy dispersal, but all sort of organisms, species, and larger taxa are disposed of as the biosphere increases its ability to access and make use of energy reserves. Thus those who criticize the notion by suggesting that organisms are exposed to hardships or "eat their own children" do not effectively challenge the notion of a planetary phenomenology that regulates itself in the manner of a body. Indeed, the research program of an environment deeply embedded in material feedbacks has been coopted by geology and geosciences departments under the name of Earth system sciences, which appropriates James Lovelock's systems cybernetic approach while distancing itself from perceived new age baggage.

Oxygen accounts for about one-fifth of our atmosphere; the mean temperature of the lower atmosphere is about 22 degrees centigrade and its pH is just over 7. Although the sun's luminosity is thought to have increased over 30 percent since the first life appeared on the planet and although combustible oxygen instantly reacts with many sorts of molecules nonetheless normally present in the atmosphere (because their concentrations are physiologically replenished), these values (temperature, pH, and the distribution of reactive gases) have remained extraordinarily stable for hundreds of millions of years. A major argument here has to do with free oxygen, which once existed only negligibly in the atmosphere: Yet, far from the increase in oxygen being a counterexample to Gaian regulation, the switch from a relatively languid anaerobic to a higher energized redox planet can also be taken to represent a planetwide metamorphosis, a violent organic homeorrhesis akin to (and in some ways perhaps even formally similar to) developmental changes such as the hormonal floods of puberty or the genetically mediated transformation of a caterpillar. (Unlike reactive machines, proactive living beings return to previous states that themselves are often changing, so-called moving set points; they are often homeorrhetic, meta-stable, or fluidly stable, rather than homeostatic, simply looping back to the same value, like a set temperature on a thermostat.) Unlike animal bodies, the living Earth—qua its proximate resources—is one of a kind and thus is not exposed to natural selection. Nevertheless, the global regulation of geophysical values over evolutionary time may be likened to the regulation of the body temperature of a mammal over decades; this allows one to speak meaningfully of a Gaian "physiology," of Earth's surface as being alive.[14]

Sociologically, Gaia ties into animism, native Americanism, and ecologism, and it provides a sort of immanent goddess that for many at last suggests a welcome departure from a transcendent God. Phenomenologically, the switchover to a Gaian worldview, to a perspective in which one inhabits not a static environment but the responsive tissues of a planet-sized complex organism, can hardly be overemphasized. The greatness of the being within which we dwell even provides an explanation of our relative ignorance of it. Gaia theory has also been threatening to philosophers and scientists, for it has occasionally served as a platform for a new age joy slide into the muck of planetary personification. If biocentrism is currently a prime grove for the culling of noble fictions, then certainly the tree of Gaia, at the very best, bears some of the most tempting fruit.

Gaia theory has been attacked on several fronts: as unscientific, "as either trivial or untestably metaphoric from the viewpoint of analytical philosophy," as an antihuman polemic, mere Green politics, industrial apologetics, and even as ecological "Satanism."[15] Yet Gaia theory thrives as a cross-disciplinary science so new the amniotic blood of the mythopoetic still adheres to its newborn skin, announcing its status and making it vulnerable to attacks from the established sciences of geology, biology, and atmospheric chemistry.

Seductive and enchanting, Gaia theory had a poetic genesis. When NASA prepared for the *Viking* mission that landed on Mars in 1976, scientists were asked to design experiments that could detect life on the red planet. Lovelock had already invented a mechanism by which minute concentrations of chlorofluorocarbons (said to disrupt the ozone layer) are detected at concentrations as scanty as a few parts per billion. Lovelock, already passionately observing the effects of life on the atmosphere, suggested that the absence of life on Mars might be detected from Earth. Earth's atmosphere, he pointed out, differs greatly from those of Mars and Venus, which are both more than 95 percent carbon dioxide—that is, stable, unreactive mixtures of gases predictable from laboratory experiments.

Earth's atmosphere, by contrast, is inherently unpredictable, containing volatile gases such as methane and hydrogen, which should not normally be found with oxygen. Lovelock reasoned that the atmosphere, far from being a sterile container for life, is inseparable from it, like the shell of a tortoise or the nest of a bird. The atmosphere is at once life's circulatory system and its skin; if life existed on Mars, its natural chemical processes would drive the Martian atmosphere away from equilibrium. But because the gases of the Martian atmosphere are in equilibrium, he argued, there was no need to go to Mars to show it was devoid of life. Needless to say, NASA, on the verge of liftoff, was not overly thrilled.

Lovelock proposed that life on Earth must have monitored its environment on a planetary scale. How else could the gases that comprise it remain in such an unstable situation? The oceans and air of Earth appear to be continuously *physiologically* stabilized, as are the body chemistry, internal temperature, salinity, and alkalinity of many organisms. Views of Earth from space, by astronauts or in kitsch postcards, have literally changed our perspective. The essayist Lewis Thomas, contrasting the Earth seen from space with the dry-as-a-bone moon, has called Earth the only "exuberant thing in this part of the cosmos," a turquoise orb with the "organized, self-contained look of a live creature, full of information, marvelously skilled in handling the sun."[16] Lovelock asked his country neighbor, the novelist William Golding, for a "good four-letter word" to express the idea that Earth has, beyond just a physical chemistry, a physiology. Golding proposed the Greek earth goddess Gaia, mother of the Titans.

The use of a mythological title to describe a serious subject for scientific study is fraught with dangers and opportunities not unlike those of

Sigmund Freud's borrowing of the Narcissus and Oedipus myths for use in elaborating psychoanalysis. Save for the Catholic Virgin Mary, mother goddesses have virtually vanished from the modern West; the unconscious association of the reappearing goddess Gaia with Mother Mary, however—if true, a kind of overcompensation for the transcendent phallocentrism of Judeo-Christianity—would help account for some of the shrillest among Gaian new ageists, whose moralistic slogans and puritanical admonitions are marshaled to save a (supposedly) pristine Earth. Yet, as Mary Catherine Bateson points out, it is of little use to treat Earth as a living body while living female bodies themselves still do not command cultural respect. Perhaps considering Earth a Narcissus-like extension of the human self works better than a neo-Christian feminization.

Although satellites had been envisioned as providing us with a view of "the whole earth," in fact, what has been provided is only one-half of the surface, the "face" of Earth. And although they depend on personification, mythology, and the forever unfinished and eroticized nature of human symbolization, such ideographic or iconic chains—Narcissus, earth, expressive surface—enhance the ethical status of Earth by giving it (a) face. The alternative is to save (this) face by considering Earth as faceless—inanimate, insensible, and unresponsive—and therefore ourselves as unaccountable to it.

It is also possible to argue that "unprovable" Gaia is a species of noble lie, a "narrative integration of cosmology and morality," and that, Western intellectual and moral biases against deception aside, Gaia is simply environmentally useful whether or not it is true. If this is the case, then describing Gaia theory as geophysiology may confer on this infant discipline the nominal equivalent of scientifically correct swaddling clothes. The culturally valuable noble lie (or ironic commentary thereon) dates back at least to Plato's advocacy of a myth of metallic origins in the *Republic.* Like the topology-violating magician who slides a ring onto a knotted loop of string, the well-told noble lie seduces the beholder into believing (in) it despite knowing better.[17]

In my book *Biospheres,* I argued that "artificial" ecosystems—containing humans, technology, and the requisite elements for long-term recycling in materially closed environments—are not all that artificial but, rather, the first in a batch of planetary propagules whose proliferation is in keeping with prior epochal evolutionary developments (e.g., bacterial spores, animal bodies, plant seeds).[18] Despite the exposure of the scientific inadequacy of the initiators of the world's biggest closed ecosystem,[19] Biosphere 2 in Oracle, Arizona, other recycling ecosystems—including one that had been planned at the world's biggest cathedral, Saint John the Divine in New York City, which would have digitally transmuted atmospheric gas measurements into music—will likely eventually be built. Already, greenhouses and buildings with central air conditioning represent a step in the direction of closed ecosystems large enough to contain human beings; the next step is the ability to completely recycle gaseous, liquid, and solid wastes.

Fully recycling self-enclosed ecosystems may first be built by the Japanese, whose limited space, island history, and technological prowess are sure to keep them interested in a project whose lucrative applications include pollution control and space station design. Farther in the future, functional biospheres may become necessary because of spoilage of Earth's shared atmosphere. I believe that the pollution-engendered, technology-fostered cropping up of biospheres (already in its initial phase) will have represented the appearance of individuality at the planetary level. This level represents a natural continuation of the microcosmic level of the prokaryote and the eukaryote made from prokaryotes and of the mesocosmic level of animal and plant bodies made from reproducing eukaryotes.

Taking his cue from Vladimir Vernadsky (who popularized the term *biosphere*), Georges Bataille writes:

> Solar radiation results in a superabundance of energy on the surface of the globe. But, first, living matter receives this energy and accumulates it within the limits given by the space that is available to it. It then radiates or squanders it, but before devoting an appreciable share to this radiation it makes maximum use of it for growth. Only the impossibility of continuing growth makes way for squander. Hence the real excess does not begin until the growth of the individual or group has reached its limits.[20]

Vernadsky was a kind of "anti-Lovelock," who, far from imagining the Earth to be alive, considered life a kind of mineral. Yet these two scientists, Vernadsky and Lovelock, are linked by their heuristic dismantling of the boundary between life and its environment, biology and geology—

what Lovelock calls academic apartheid. But from either perspective, the appearance of materially closed ecosystems capable of recycling carbon, nitrogen, sulfur, phosphorus, oxygen, and other elements necessary to a total system of life, including human life, merits attention. Supererogatory biospheres, in Bataille's schema, extend the limits of growth. Within that general economy beginning with solar radiation and bacterial photosynthesis, "bonsai" biospheres make use of the solar-driven material surplus generated by what Vernadsky has called the pressure of life. Bonsai biospheres funnel into a new form of life the metabolic reserve whose lavish squandering Bataille has described as a fundamental feature of cultures less acquisitive and profit-oriented than our own. The appearance of closed ecosystems within the general planetary ecosystem makes clear that the biosphere has a fearful symmetry of its own.

BACTERIAL OMNISEXUALITY

Bacterial omnisexuality refers to the genetic exchanges among bacteria considered promiscuous in the sense that they do not delimit these exchanges with species barriers.[21] Theoretically, any bacterium can at any time in its life cycle give a variable quantity (rather than exactly half, as occurs during the meiotic sex of plants and animals) of its genes to any other bacterium, although it may require intermediaries such as plasmids or viruses to do so. Bacterial omnisexuality was the first type of sex to appear on the planet, some three billion years ago. It was always crucial to the biota's ability to react quickly to environmental changes and emergencies, since the *lateral* transfer of useful traits among rapidly reproducing bacteria is of far greater environmental consequence than the slow, vertical inheritance of meiotically reproducing organisms. (Although bacteria are termed asexual, this refers only to their means of reproduction.)

After the evolution of eukaryotic cells from symbiotic bacteria, bacteria omnisexuality became important as a way of genetically "locking" together once diverse groups of prokaryotes. We are accustomed to thinking of them merely as germs, but most bacteria are harmless to humans. Bacteria are biochemically and metabolically far more diverse than all plants and animals put together. The natural history of bacteria is so bizarre that they would have excited huge interest had they been discovered in outer space rather than beneath our feet. All things considered, bacteria appear crucial to the upkeep of Gaian systems of sensation, feedback, and physiological control. Indeed, Gaia may have appeared on Earth some three billion years ago basically as an emergent phenomenon of bacterial crowding.

Four-fifths of the history of life on Earth has been solely a bacterial phenomenon. Moreover, all plants, animals, fungi, and the miscellaneous eukaryotic kingdom known as protoctists are bacterial in nature. The nucleated, mitochondria-containing eukaryotic cell on which all non-bacterial forms of life are modularly based is itself the result of symbiosis and bacteria recombination (omnisexuality). The xenic origins of the eukaryotic cell have major implications for the self, the body, and a vulgar Darwinism that equates evolutionary success with competition. With respect to the bacterial colonization prerequisite to Gaia and its global metabolism, animals including humans are epiphenomenal. There seems little doubt that even full-scale nuclear war could not destroy the bacterial infrastructure.

Eukaryotic cells evolved through a process known as endosymbiosis. Perhaps the simplest model of endosymbiosis is for one organism to swallow another without digesting it. In microbes especially, because of their lack of an immune system, organisms may be eaten that are likely to survive within their hosts. A more complex form of endosymbiosis is bacterial infection: in this case, too, death does not ensue but, rather, the invading organisms successfully reproduce inside, and in some cases may have become absolutely required by, their hosts.[22]

Not only the origin of new species but the origin of the metakingdom Eukaryotae as well, comprising all nonbacterial organisms, occurred not through gradual accumulation of mutations but through endosymbiosis; we may owe our very existence to the ancient "failure" of Lilliputian vampires, oxygen-respiring bacteria similar to modern-day *Bdellovibrio,* to kill the hosts whose bodies they had invaded. This was of course a Pyrrhic victory, since these organelles now energize our entire bodies. They are now generally well behaved, although cancer is noteworthy as the enhanced activity of mitochondria, which provide energy for tumor growth.[23]

Technogenetic manipulation of bacterial strains, which promises huge financial returns from the biomedical market and, ultimately, a radical refashioning of the human genome into new species, *is* bacterial omnisexuality—bacterial omnisexuality ministered, "engineered" by human hands. If eukaryotes could trade genes as fluidly as do bacteria, it would be a small matter for dandelions to sprout butterfly wings, collide with a bee, exchange genes again, and soon be seeing with compound insect eyes. Bacteria are able to trade variable quantities of genes with virtually no regard for species barriers. Indeed, despite a lingering Linnaean nomenclature, bacteria are so genetically promiscuous, their bodies are so genetically open, that the very concept of species presents us a false conceptual shackle falsifying the genetic fluidity of these ancestral life-forms.

Bacteria are omnisexual. Genes received by bacteria in one generation are passed down indefinitely thereafter during cell divisions. The discovery that most of the DNA in the genomes of eukaryotic organisms is "redundant," coding for no known proteins, suggests that it may be left over from the merging of stranger bacteria whose incorporation produced superogatory information, genetic "deadwood." An example of bacterial recombination is the evolution of penicillin-resistant staphylococci. The gene that directs the synthesis of an enzyme that digests penicillin probably arose in soil bacteria. But via phage-mediated omnisexual exchanges, staphylococci have incorporated such resistance and survived the hospitalization of their hosts. Omnisexuality makes bacterial boundaries plastic and forces us to view bacterial cells not in isolation but as the cell of an extremely diffuse yet continuous Gaian body. Indeed, Sorin Sonea has postulated that such horizontal gene transfer among bacteria qualifies them as a single superorganism whose body coincides with the surface of the planet.[24]

Aside from fictions of Gaians using bacterial omnisexuality to remodel their bodies after the image of beauty or strength or even the demihuman metazoans of Greek myth, where does the confluence of bacterial omnisexuality and evolving notions of the human body lie? Whether discussing the disappearing membranes of endosymbiotic bacteria on their way to becoming membrane-bound organelles, or the current changes within the global human socius, the rectilinear notion of the human self, the bounded, stands challenged today from yet another viewpoint, that of the new biology. This zoological "I" is open to radical revision.[25]

How does a concept of the individual that leans toward the physical model of bacterial omnisexuality and aesthetic model of a *différance* differ from the "encased self" model of zoocentrism? One example is that used by Burgess—the artist whose production, genius, or gift results not from her or his own body but from the interference patterns generated by a series of symbiotically living forms (spirochetes in this case). The disease that causes discomfort and near madness is also a symptom of a musical disturbance of former ecological harmony, of what was once environment, *oikos,* but is now neither home *to* nor home *of* but rather body. As an organism's connections to the external environment grow, that environment becomes its body. Like the snail whose house is carried on its back, the "case" of the "self" has been moved, through an incorporation of what once would have been called inanimate matter—admittedly, organically worked and reworked. The boundaries of selfhood are expanding. In microbial ecology, the "I" is literally a figure of large numbers. Pieces of the self—from plasmid and viruses to laboratory-spliced genes and prostheses, from milking machines to mechanical and real hearts—are obvious examples of a circulation of elements of subjective identities always already undergoing active (de)composition. Because the self is not closed but open—for the relations of the elements of physiological identity and psychological subjectivity link up with all matter through all time—it would be hasty to dismiss the general medieval scheme of microcosmic correspondence as mere superstition. Nor, of course, is this in any way to suggest a one-to-one linkage or reliably complete mapping for the series prokaryote–protist–person(a)–planet.

Today, for humans, the body and the self are most clearly in a state of fundamental Heraclitean change. The proverbial river is recognized as a conduit in the circulatory system of a being that has exerted control over the composition and redox state of its atmosphere for hundreds of millions of years.[26] Ostensibly, human bodies are integrating newly evolved and evolving viruses, only some of them, such as herpesvirus, identifiable from their pathogenicity. The majority of viruses and bacteria circulate around the biosphere and technosphere harmlessly and unnoticed, joining together fragments in *jamais-vu* combinations. Humans, too, are not merely *zoe* or metazoa, in the sense of bare life, or mitotically cloned cells differentiating from an embryo into tissues. We are metametazoa, metazoans whose industrial pollutants, ecological impact, and telecommunications have not only altered the shape of life on Earth but forced us to recognize the environment of the sum total of life on Earth as a totality with shared destiny, as a single, integrated, sensitive, and sensing system.

Life, according to my mother, is *bacterial.* And this bacterial world, according to Lovelock, has a life span. The biggest challenge to life over the long run has little to do with the paltry meanderings of human beings. It comes rather from the source of all life, the Sun. According to astronomical calculations, the core of the Sun is expected to swell as helium begins to fuse with carbon in nuclear reactions, luminosity increases, and the Sun becomes a "red giant." To forestall a dangerous heating of Earth attendant with the death throes and rise in temperature and expansion of the Sun would likely involve carbon dioxide. As is widely known, carbon dioxide is a "greenhouse gas" whose presence in the atmosphere heats Earth by trapping infrared radiation. Gaian scientists believe that (over the long run) life has managed to sequester increasing amounts of carbon dioxide from the atmosphere to counter the effects of a Sun that has been growing steadily more luminous since the inception of life on Earth. The carbon dioxide that has vanished from Earth's atmosphere exists on the terrestrial surface in the form of carbon-containing minerals and carbon-based life-forms. If the biosphere has indeed been removing CO_2 keeping itself cool, Gaia's future as a terrestrial being extends only some hundred million years: there is only so much carbon dioxide that can be removed from the atmosphere to counter the increasing luminosity of the Sun (and this, of course, assumes a total reversal of the recent increase in atmospheric CO_2, because of human industry).[27]

Although imminent from a geological point of view, a hundred million years is about twenty times the average life span of a vertebrate species. It is almost certain that by this time *Homo sapiens* will have become extinct or speciated. Humanity as a species is no more distinct than animals as individuals, and I have tried in this essay to use the new biology to relativize that zoocentric bedrock, the bounded, autonomous self. In a certain sense, this relativizing represents a preliminary sacrifice to the Sun, whose red giantism "we" at least will have escaped.

CHAPTER 13

KERMITRONICS

The choral song which rises from all elements and all angels, is a voluntary obedience, a necessitated freedom. Man is made of the same atoms as the world is, he shares the same impressions, predispositions, and destiny. When his mind is illuminated, when his heart is kind, he throws himself joyfully into the sublime order, and does, with knowledge, what the stones do by structure.
–*Ralph Waldo Emerson,* The Conduct of Life

ker·mi·tron·ics
noun pl [ker-mi-*tron*-iks]
singular in construction: the philosophical understanding and application of ourselves as sensuous and responsive beings not, however, under our own control. First used 2012. From Kermit the Frog, a Muppet (= combination of marionette and puppet), first introduced in 1955.

LIKE THE MIME IN A CIRCUS who pretends, from the dirty floor, to balance the high-wire walker, or the clown who, twirling her fingers with a gleeful simper, seems to send the acrobats falling head over heels in their aerial somersaults, before reaching through thin air to catch a helping hand, so we may be pretending that we are running the show. Only in our case we don't seem to know we are pretending. The performers wear no special clothes. They are not paid union dues or the celebratory object of a special occasion. The big top, far from being the circumscribed arena of a circus tent, is the surprisingly vast if more intimate space of our own head.

The fact that, under experimental observation, tiny foraminifera select certain sizes and colors of glass beads to build their living skeletons, shows that even these tiny creatures have choice. But is it free? Doesn't a computer have choice as it goes through the complex sifting, winnowing process of its electronic operations?

In 2002 I impressed undergraduate students in an English class in Danville, Kentucky, who'd read my cowritten work (with the plagiarized title) *What Is Life?*, in which life is defined as "matter that chooses." I know that they read the book because they looked genuinely surprised when I admitted I no longer knew if I believed in free will. Three and a half centuries before recent experiments that seem to prove him right, the philosopher Benedict de (or, in his Jewish appellation, Baruch) Spinoza argued, on the basis of logic and intuition, that we do not really have free will. Rather, we are simply not aware of the causes of our own thoughts and behavior. Observing those thoughts and behavior, we jump to the conclusion that we caused them. And, unlike that clown, who entertains us but is not fooled by her pretending to control the trapeze artists, we fall for it, again and again, spellbound by a show that seems to work itself, with neither clown nor safety net.

"So the infant," Spinoza writes, "believes that he freely wants the milk; the angry boy that he wants vengeance; and the timid, flight. Again, the drunk believes it is from a free decision of the mind that he says those things which afterward, when sober, he wishes he had not said. Similarly, the madman, the chatterbox, and a great many people of this kind believe that they act from a free decision of the mind, and not that they are carried away by impulse. Because this prejudice is innate in all men, they are not easily freed from it."

Lift the book, put it down. Decide, decide not to decide.

Although now I can see that "choosing" does not necessarily mean "freely choosing," I might have to change my mind again, whether I choose to, sensu stricto, or not. There is no question that we feel we are free, up to a point, but there is no proof for it either: we may just be unaware, as Spinoza clearly explained three centuries ago, of the causes of our behavior. It is what my mathematician friend Steve Shavel calls "the last Ptolemeic." Like the idea that we are the center of the universe (as we were in Ptolemeian astronomy), the idea that we are the originators of our own thoughts and behaviors is one we humans don't want to give up.

Indeed, it is an insult. I can look out my window and *see* the Sun come up in the east. In the same way I can *feel* that it is my free decision to choose this very specific and, hell, sesquipedalian word. How dare Spinoza, let alone Sam Harris three centuries after the fact, call into question the subjectively obvious reality of my own freedom.

I remember my astonishment upon visiting my father's house above the gorges of Lake Cayuga in Ithaca, sitting on his plush black leather couch before a glass coffee table in an open room and "proving" my freedom by lifting a water glass of my own volition.

Although the discussions I had with him sometimes hid our emotional attachment and conflicts under the umbrella of intellectuality, this particular conversation seemed to be all, or mostly, business. I was in the company of a red-headed girlfriend who was a great admirer of science and thrilled to meet my famous father. I was feeling good about myself not only because I was with her but because I had just finished a book, *What Is Life?* that, unlike some of my other attempts, finally seemed to be a successful blend of science and Continental philosophy. Influenced by Heidegger's student, the philosopher of biology and scholar of Gnosticism, Hans Jonas, the book made a strong argument that one of life's signal traits was its sentience and, part and parcel of that, its existential power over itself, its power to choose.

The friend took what I then thought was the underdog position in her discussions with me. Determinism was such a powerful scientific principle, and the causality of chemical and physical links so powerfully explanatory throughout nature, that there did not seem, that we had no call—other than the usual hubris—to exempt our selves from these causal networks. If I thought that I, like Samuel Johnson rebutting Bishop Berkeley's idealism with the kick of a real rock, was going to lay to rest scientific determinism simply by picking up a glass and saying, "I decide," I had another thing coming to me.

Just because it seems that Earth is flat and at the center of the universe, just because it is comforting to think that I am controlling my actions, doesn't make it so.

Niels Bohr, who argued with Albert Einstein's famous determinism, going so far as to say that Einstein should quit telling God what to do, also said, "There are trivial truths and the great truths. The opposite of a trivial truth is plainly false. The opposite of a great truth is also true."

Can free will and its opposite both be true? Another great physicist, Erwin Schrödinger, seemed to think so:

> According to the evidence put forward in the preceding pages the space-time events in the body of a living being which correspond to the activity of its mind, to its self-conscious or any other actions, are, if not strictly deterministic, at any rate statistico-deterministic. For the sake of argument, let me regard this as a fact, as I believe every unbiased biologist would, if there were not the well known, unpleasant feeling about "declaring one's self a pure mechanism." For it is deemed to contradict free will as warranted by direct introspection. So let us see whether we can draw the non-contradictory conclusion from the following two premises. . . . My body functions as a pure mechanism according to the Laws of Nature. . . . Yet I know by incontrovertible experience that I am directing its motions of which I foresee the effects, that may be fateful and all-important, in which case I feel and take full responsibility. The only possible inference from these two facts is I think that I—I in the widest meaning of the word, that is to say, every conscious mind that has ever said or felt "I"—am the person, if any who controls the "motion of the atoms" according to the Laws of Nature. . . . It is daring to give to this conclusion the simple wording that it requires. In Christian terminology to say "Hence I am God Almighty" sounds both blasphemous and lunatic. But please disregard these connotations for the moment and consider whether the above inference is not the closest a biologist can get to proving God and immortality at one stroke.[1]

I think we may be able to make sense of this seemingly bizarre statement if we consider time. The past seems solid to us, impermeable, closed. The future, however, the now on which we are on the edge, upon which I am as I hesitate midstream in the finishing of this sentence, seems open, liquid—not closed but close, alterable by thought, permeable by whim.

But can this be so? The past was once the present; they are part of the same process . . . so maybe the difference we see is an illusion, somehow to do with time. If we were raised up to another dimension, then maybe the differences between the great truth of freely deciding (if only to a minor degree our own movements and cable channels and friends and foods) and being cryptically forced to do things would be rectified. The mystical *coincidentia oppositorum,* the unity of opposites, would be realized—not by a logical contradiction but by a shift of perspective.

It would not be the first time appearances were found wanting. Any number of optical illusions show that shapes and sizes can differ from our perception; Benham's wheel, with only black and white curved stripes, produces light browns, reds, and greens that quickly vanish when it stops rotating. Because of the equidistance of our ears, we easily can mistake a sound in front or directly on top of us for one behind us.

But in general we realize we have been fooled only when one or a number of senses contradict and correct the initial illusion. Thus, if you insert your pencil into a glass of water, it will seem to be bent because light refracts it. To prove this isn't so you only have to touch the pencil or remove it from the water. The separate senses provide redundancy, giving us a generally coherent picture of the world and allowing us to revise anomalies when they arise. When we can't, as when a magician does a sleight of hand, we may conclude that our senses, perhaps purposefully misdirected, have misled us.

Giuseppe Trautteur, a physicist at the University of Naples, is interested in what he calls the "double feel" of free will. On the basis of logic and experimental evidence, Trautteur accepts that free will is an illusion, but he argues that it differs from other illusions because, unlike them, it cannot be directly tested. For example, we check with a ruler that the two lines in the Müller-Lyer illusion are actually equidistant, or with our hands that the sight of a pencil apparently bent in a glass of water is in fact an optical illusion. There is no such recourse to redress the strong feeling of free will that, however, we can learn to experience doubly. Experientially it is there, but cognitively, as we investigate it, it begins to fade. Trautteur calls "double feel" this experience "of a conscious subject who feels he is performing a free action and at the same time is ineluctably convinced of the illusory nature of free will."[2]

It is interesting to speculate how Spinoza, who seems to have been the first person to have logically deconstructed free will, came to his conclusion. Deeply influenced by René Descartes, Spinoza turned the dualist on his head. He was what I would call a Cartesian monist: someone who accepted Descartes's influential separation of reality into "thinking stuff" (*res cogitans*) and "extended reality" (*res extensa*), but then put the two—mind and matter—back together again. The speculative answer to my question is rather obvious: in recombining the artificially separated mind from matter, Spinoza returned to it the attribute science had granted matter, considered causal, "mechanical," explicable by laws of physics. Reunified with matter, mind, including God's mind and humanity's, became as causal as the motions of the planets or the eternally valid relationship of two lines.

Descartes, you'll recall, who was into anatomy, speculated that the pineal gland, at the time known only from human brains, to be a sort of hotline to God and free will. For Descartes, humanity shared in the divine ability to choose outside the causal network of science. The world of things was separated from the world of thoughts, which we alone possessed in the animal kingdom. It excused our use of them, unfeeling brutes, investigable automatons. It also allowed us to examine the material world as a mechanism. This seems to have been a sort of compromise with the church, concerned about the increasing successes of the scientific method and its encroachment into the sacred realms of the heavens and life.

Spinoza's monism owed much to Descartes but was more consistent—and it had far-reaching political and scientific repercussions. I knew Spinoza was a huge influence on science: he was a great favorite of Einstein and I remembered that my father, an atheist, said he believed in the God of Einstein and Spinoza—which is, in a sense, to say that he did not believe in God at all, because, depending on how you read him, by "God" Spinoza means Nature—not just visible Nature but nature as timeless and infinite, the impersonal, certain, necessary, eternal, and true realm accessed by the mathematicians and geometers. Spinoza extended Descartes's realm of geometry and necessity to include the human mind and God. He basically took back what Descartes had granted to the church. There were no special allowances where the realm of necessary relationships and causality did not apply. In fairness, Descartes, having learned from the examples of the tortured Bruno and the imprisoned Galileo, was trying to save his ass if not his soul. Spinoza, who translated into Dutch one of Descartes's works, was probably not as worried. He also may have been braver. By evening out the playing field, by removing the get-out-of-necessity-free card that Descartes (smoothing over potential problems with the church, by avowing an exceptional status to humans) reserved for people, Spinoza collapsed the Cartesian dualism that still affects us, between the *res extensa* and the *res cogitans,* material and cognitive reality. In doing so, he unified a world that had arguably been illicitly split. Which is why I think it's fair to speak of him as a Cartesian monist. He was playing the same game as Descartes, but he got rid of the special rules, the anthropocentric cheating!

Einstein loved Spinoza. He followed him in saying he believed in a God that was impersonal, unchangeable, unconcerned with human affairs. Einstein's famous statement that the Old One (God) does not play dice, which he wrote to his physicist friend Max Born in a letter, also testifies to his Spinozism.

Einstein was such a fan of Spinoza, whose library he visited on November 2, 1920, that in Einstein's effects there is a partly crossed-out poem titled "To Spinoza's Ethics," which reads in part, "How I love that noble man / More than I can say with words," and ends, "You think his example shows us / What human teaching has to give / Don't trust the comforting mirage: / You have to be born to the heights."[3]

Maybe poetry is, as Robert Frost said, what gets lost in translation, but it does not seem to be a very good poem, not as poetic, certainly, as Spinoza's view of an endless, purposeless, and uncreated universe, with infinite dimensions.

Like Schrödinger, Spinoza had his own *coincidentia oppositorum* with regard to freedom. Born to Jewish parents in Amsterdam, Spinoza (1632–1677) came from a family who fled to Portugal from the Spanish Inquisition and then fled Portugal when the Inquisition came there. Spinoza's grandfather Isaac brought the family to Nantes in France, and they were again expelled in 1615. Compounding these experiences of intolerance by the Catholics, Spinoza dealt with the intolerance of the Jews: he was excommunicated by the Amsterdam synagogue and declined an academic appointment at Heidelberg, which led him to live more humbly as a lens grinder, a job that afforded the luxury of the ironic privilege of thinking freely on the beautiful harmony of a universe that was limitless in its determinism. Spinoza's *Theologico-Philosophical Treatise,* written during a hiatus in the writing of his philosophical masterwork, *The*

Ethics, and precipitated by the death of an imprisoned friend, helped lay the political foundation for separation of church and state, as well as the freedom of speech and worship.

John Locke, who was a big influence on America's founding fathers, not only read all of Spinoza but, born in the same year as Spinoza and living for a time in the Netherlands, also read all the books Spinoza mentioned. I knew Spinoza was crucial for science and philosophy. He seems, for example, to be one of the only major philosophers Friedrich Nietzsche knew of and didn't lay into. But I was not aware of Spinoza's importance for politics. It seems to me remarkable that Spinoza, the high priest, if you will, of causality, who took such pains to disabuse us of our tendency to want to grant a special freedom to our thinking process, was also a great political defender of, you guessed it, freedom: freedom of speech, freedom of thought, freedom of religious worship, and freedom of assembly. On the one hand, Spinoza denies we are truly making our own motions. We do not control the motions that Schrödinger says make us gods.

"Further conceive," says Spinoza, "I beg, that a stone, while continuing in motion, should be capable of thinking and knowing, that it is endeavoring, as far as it can, to continue to move. Such a stone, being conscious merely of its own endeavor and not at all indifferent, would believe itself to be completely free, and would think that it continued in motion solely because of its own wish. This is that human freedom, which all boast that they possess, and which consists solely in the fact, that men are conscious of their own desire, but are ignorant of the causes whereby that desire has been determined." On the other hand, he was so committed to the cause of political freedom that he interrupted work on *The Ethics* to write the *Theologico-Philosophical Treatise,* a spirited defense of civil liberties that informed the founding documents of the United States.

Spinoza seems to have managed the difficult defense of freedom and its opposite by reinterpreting freedom as a kind of purposeful acquisition of knowledge. We might not control it, but we feel we do and are genuinely changed—for the better—by its effects. Learning where and what and who and how—and now we might add (courtesy of paleobiology and nonequilibrium thermodynamics) when and why we are—is the very motor of our happiness. Happiness's pursuit is the pursuit of knowledge. This is philosophy, from Greek words for "love" (*philos*) and "wisdom" (*Sophia,* its goddess).

If the philosophical consistency of Spinoza and the scientific success of his fans like Einstein and my dad are enough to unsettle our conviction of being rare cosmic repositories for freedom, experimental evidence hasn't helped matters much. Connecting brains to scans, scientists recorded when a person felt he or she had decided to move. In all cases, significant brain activity in the motor cortex was identified *before* the time participants thought they made a decision. On average they took 0.2 seconds between deciding to push a button and actually pushing it. Nonetheless, their subjective experience of detecting an *urge to act* in fact *followed* associated electrical signals in the secondary motor cortex by up to 0.3 milliseconds. To make a long story short—apparently a version of what we do when we think we think we are deciding—there is about a half-second delay between when we feel we've decided and when "our brain" decides. Half a second before "we" burst out the starting gate, the brain has already neurologically fired. We never jump the gun. The brain's neurologically quicker, beating us to the punch, always.

Other experiments show that lateralized brain hemisphere activity prior to hand movement allows for roughly 70 percent predictability as to whether a subject will later "decide" to raise his or her left or right hand. Still other experiments suggest that both subjective identification of a conscious intention to move and the movement itself follow preliminary movement and unconscious brain activity by about 2.8 seconds.[4] As research continues, the suspicion is growing that our intention to move is not a cause but a sensation of bodily movements that have already taken place.

In the 1990s, before he started dating the determinist redhead who admired my father and is mentioned above, my brother had an e-mail tag that ran, "I have a strong will and a weak won't." I'm not sure if he intended it as a lure to women or a warning, but apparently it did the trick. The experimental evidence strongly suggests that all of us have strong wills and weak won'ts, and are no match against the quick-draw of our ivory-handled brains. It strongly suggests that we cannot stop or deflect those bullets because we are those bullets.

And if free will is an illusion, there may be repercussions. Why do some behaviors land us in prison while others confine us to the hospital?

Samuel Butler, in his strange morality novel *Erewhon,* tells of people who are imprisoned for typhoid fever, but who are coddled sympathetically by their friends when they come down with a propensity to commit a minor crime, for example, when a woman suffers an outbreak of shoplifting. Elsewhere Butler argues that microbes make little choices that, over time, become habits such that, for example, we no longer remember when our ancestors first decided to grow an eye. The historical background for these literary-philosophical experiments comes from his disenchantment, after great initial enthusiasm, with Charles Darwin's portrayal of evolution as largely mechanical. Given that unconscious processes are so complex, how can we distinguish between when "we" decide to do something and when something happens automatically? Between what we can help and should be responsible for, and what we can't help and is not our fault?

Trautteur, at the end of his essay on the double feel, wonders what his colleagues are thinking when, after adducing all sorts of good arguments against free will, announce that they believe in it anyway, or that they don't but it doesn't matter. Doesn't it? Aren't they tempted to rob a bank and take off for a South Seas island, given that they have no choice?

Further experimental evidence suggests that, irrespective of whether free will exists or not, consequences accrue to those who believe in it. Think about what you are thinking right now. Who is thinking it? Where is it coming from?

Consider the words and images that shadow forth and turn into water scenes as you doze. Are "you" the origin of them, or are you simply their observer? And where are you, who are you, if you are not those thoughts or words but just their observer?

It thinks, wrote Nietzsche. You are at a reality carnival, on a ride. No clown is moving the trapeze artists of your thoughts. They are just doing their own thing.

It has yet to be proved that anything could be otherwise. The wave function or sum over probabilities of quantum physics may suggest this, but that still does not seem enough to rescue the "last Ptolemeic." Neither quantum probabilism nor garden-variety determinism, the complexities and algorithmic surprises of deterministic chaos that are neither predictable nor free, can save free will. Perhaps only a radical leap of faith can save free will. "I believe in free will because that is my choice."

Notice I said "because." Free will exists because I say so. Free will exists because it has no choice.

Francis Crick, codiscoverer of the molecular structure of DNA, also became interested in free will. He suggested that our feeling of free will is based on a part of the brain, the anterior cingulate sulcus. I think it would be funny if we located, long after Descartes, the part of the brain that controls free will. It reminds me of a comedy skit. A great scientific discovery is made, and an even greater press conference is called. In dour tones in one of the studios of one the most popular television networks, an announcer interviews a scientist who announces that at long last the elusive human sense of humor has been located. After a commercial break for adult diapers, it is revealed that it resides in an amplified ionic pattern in the membrane of the endoplasmic reticulum.

In fairness, Crick postulated it was our sense of free will, not free will itself, that had been located. If free will doesn't exist, it will be difficult to locate it.

What then?

The press conference is over, leaving us hanging, and now there is a show on the prehistory of muppetry. An expert discusses the German romantic writer Heinrich von Kleist (1777–1811). Kleist glossed the perplexing agency of puppets, which seem to come alive at the end of their strings, and whose mechanism, one of sensitivity to their surroundings combined with a kind of ease of movement, allowed them to dance as well as, if not better than, their handlers, their would-be puppeteer-controllers:

> One evening in the winter of 1801 I met an old friend in a public park. He had recently been appointed principal dancer at the local theatre and was enjoying immense popularity with the audiences. I told him I had been surprised to see him more than once at the marionette theatre which had been put up in the market-place to entertain the public with dramatic burlesques interspersed with song and dance. He assured me that the mute gestures of these puppets gave him much satisfaction and told me bluntly that any dancer who wished to perfect his art could learn a lot from them.

Sitting down—I will lower the textual marionette of Kleist now—the writer inquires further into his friend's "remarkable assertion."

"A group of four peasants dancing the rondo in quick time couldn't have been painted more delicately by Teniers," Kleist tells us, admitting to his friend the gracefulness of the dance movements of the puppets, "particularly the smaller ones."

> I inquired about the mechanism. . . . I wanted to know how it is possible, without having a maze of strings attached to one's fingers, to move the separate limbs and extremities in the rhythm of the dance. His answer was that I must not imagine each limb as being individually positioned and moved by the operator in the various phases of the dance. Each movement, he told me, has its centre of gravity; it is enough to control this within the puppet. The limbs, which are only pendulums, then follow mechanically of their own accord, without further help. He added that this movement is very simple. When the centre of gravity is moved in a straight line, the limbs describe curves. Often shaken in a purely haphazard way, the puppet falls into a kind of rhythmic movement which resembles dance.
>
> This observation seemed to me to throw some light at last on the enjoyment he said he got from the marionette theatre, but I was far from guessing the inferences he would draw from it later.[5]

The Estonian-born biologist Jakob von Uexküll coined the term *Umwelt*, meaning life-world. All organisms have such worlds, he argues, and in his writings he imagines what it's like to be a blind, deaf tick smelling the butyric acid of a succulent mammal, jumping onto a sweet patch of skin with blood beneath the surface on which to feed. He imagines (and in his essay, "A Foray into the Worlds of Humans and Animals," depicts!) the view a scallop might see of a European street. Dogs, he points out, navigate differently through space, because their four legs carry them up inclines without missing a beat and see indoor objects primarily in terms of how they may sit on them. What's it like to be a heron fishing with an iridescent beetle, a swarm of bees in a field of ultraviolet-patterned flowers? The Jesus lizard walking on water, the pebble toad, *Oreophrynella nigra*, ricocheting down a Venezuelan mountain, the queen bee laying a thousand eggs in her fragrant maze, the female mosquito finally draining blood from a host during a muggy dusk, and perhaps cells in that host, and perhaps water lilies in the pond where dives the fat bullfrog whose croak filled the sounds of the rustling forest the night before, and perhaps even the forest itself, in its slow chemically complex growth, its subterranean nuclei-trading networks of mycelial tendrils, its bacterially diverse metabolism, able to breathe arsenic and sulfur, to grow and merge, the biological world as a vast colossus of differentiable and merging sensitivities. . . . All of these and each of us, one might argue, to *whatever* extent we actually control our actions, are spectators at the greatest show on Earth. Each of us is, in a way, like an instrument in a billion-year-old symphony, making vibrations and emotions and Uexküllian music, for ourselves and one another if not, as the post-Kantian biologist may have thought, as localized beings in the perception of a multidimensional immanent being seeing itself, at least partly, through its manifold perceivers. Here *organism,* the word, returns to its root, *organon,* Greek for instrument. We play, or are played, by the manifold all.

A cartoon that whipped around the Internet in 2011 seems to have originated, proximately at least, in Japan. It shows Kermit, the famous Sesame Street Muppet and bright green frog, known for his bulging bug eyes and Robin Hood demeanor, replete with swashbuckling throat kerchief and ability to look dead serious despite the fingers that form his forehead and the thumb that articulates his lower jaw.

Kermit is consulting with a doctor. He is attentive and looks as if he could be worried. The doctor is looking at an X-ray that, because it is in the picture's foreground (we look over the doctor's shoulder), is almost as big as live Kermit on the other side of the desk, the white sheets hanging on the aluminum rod behind him suggesting that he may just have dressed or undressed.

"Sit down," says the Doctor. "What I'm about to tell you may change your life."

But there is not a seat for Kermit to sit on, and, since it is a still cartoon, no movement of any kind.

The X-ray looks accurate: It depicts the amphibian's bones, and there does seem to be a kind of problem. Instead of a normal frog skull, we see an ulna and radius, carpals and metacarpals and phalanges. That is, we see the bones of a human hand.

If Kermit were to find out, he might be quite upset.

There is not much more of an insult to one's feeling of independence to find out that one is, literally, a Muppet.

And that is the kermitronic predicament. Can we handle the letdown of not being in control? After losing the center of the universe, and having our obvious position as the mightiest, most intelligent, and attractive of species impugned by deluded biologists?

Hell, we've come this far, why not. It is not true that we are not in control, that we are being played. We are playing. We are playing ourselves in that we think we are in control. Kermitronics—or "kerm" as you might call it—is the notion that even if we do not control the stochastic-deterministic universe, we can enjoy the post-Uexküllian Spinozistic ride. It is the notion that everything is connected, so volition truly makes sense only at the level of the whole. But since this universe is infinite in extension and consciousness, as well, possibly as time, there is neither planning for the future nor the sort of feeling of free will that accompanies a human creature constrained by linear time. Call it God or the Universe, there is no free will (or purpose) at the level of the whole. At that level there is no possibility of local freedom. Everything is already always geometrically and eternally articulated. And at our level, there is no fully separable agency, so we are part of this whole.

It has yet to be proved that anything could have possibly been done different.

Got kerm?

kerm
abbr: kermitronics

CHAPTER 14

ON DOYLE ON DRUGS

I GREW UP in Timothy Leary's old neighborhood. Newton Center in the mid-1970s was past the glory days of Orange Sunshine, but a few kids knew about it. We did all right though, with our Blotter, Microdot, and Windowpane, which catapulted me, one fine afternoon, after a whole hit and an emergency purchase with shaky fingers (I was not a smoker) of a pack of Marlboros, and a harrowing walk that turned into a run, from Murray Road, the oldest alternative high school east of the Mississippi, a converted elementary school with a Ping-Pong table in the front entrance across the street from which in a swampy outback where we had tried unsuccessfully with our thumbnails to slice the tiny cellophane-like tab, I ran from the newly iridescent white ball and my friends, the older one of whom had told us to just take a whole one and, smoking with us, that marijuana always helped him come down and that no one was going to die or anything, with that word *die* in my mind, newly transfixed and pointing toward me, I ran from them (who would later come to my house looking for me and hear blaring what they thought was my stepfather yelling at me), miles along Commonwealth Avenue, wild-eyed thinking about how I could understand how Art Linkletter's daughter committed suicide, seeing now a street sign called HOMER, which took on a veneer of cosmic significance, the inside becoming the outside as it renewed me in my efforts to get back, home I must get home, until a cab driver friend of the older boy who told us it was a good idea to take a whole square of eight-way Windowpane, recognizing me and asking me if I needed a ride where was I going and I said Home, 106 Gibbs Street, inside the green house on top of which I now found myself blasting the television and being freaked out by an ad with a female model with nail polish on her claws, that not helping I fixed myself some chocolate milk, spilling it on the counter beneath the blaring radio I had turned on with my shaky hand, another misguided attempt at comfort, to run to my room. Hours later, after shaking on my single bed near the multicolored peace and love logos I'd affixed to the black-painted wall of my teenage attic room, unable to distinguish among my senses, unable to tell if my eyes were open or closed, looking through the ceiling directly into a cosmic abyss that I would later describe as eternally recurring metaphysical evil, I found myself on our front porch asking my brother for a cigarette. Calm now from all the shaking, not having expected to ever return, the experience of being outside of time so unlike any other, I now, incredibly, believed myself when I told my brother that I had not taken anything, explaining I had just taken a nap! It took me days to piece together my whereabouts during the trip and, despite all the running I'm not sure I ever did arrive there, that house where we had earlier watched the moon landing in the dining room, not quite, not as I left it.

IF YOU BELIEVE in the genius loci, the spirit of the place, then you could make the case that my friends and I were raised in the bosom of psychedelia, or in one of its other erogenous zones, in suburban Boston across the Charles River from the university, Harvard, where the West Point military academy graduate Leary had gone from straight-laced psych professor to free-love guru and self-described high priest teaching the gospel of turning on, focusing on the inner world, which can be infinite compared with the limited past and future, and dropping out of school.

But while Leary found it necessary, or at least easy, to cut his ties with repressed academia, basking in the cult of personality and the women and freedom his counterculture fame afforded, others did not sever their ties with academia and its more rigorous if conservative protocols. They pushed forward with the goals of showing the value—medicinal, cultural, and biological—of "psychedelic" drugs. One of these is Rich Doyle, a professor of rhetoric at the University of Pennsylvania, who does an Aztec two-step, performing the seemingly impossible feat of keeping one foot in academia's ivory tower and the other in the jungle. Doyle's investigations chronicle experimenters while relaying his own experiences and what they mean, or don't. This drop of the academic pose of objectivity paradoxically makes him more trustworthy, a better guide to this rugged psychic terrain than those who cling to the conceit of being credible tour guides of a country they claim never to have visited.

Doyle follows in the tradition of the early Leary before he went off into the hinterlands of the counterculture, and of Ralph Metzner, Leary's sometime coauthor, who wrote, "Those who are ideologically committed to the still-prevailing Newtonian-Cartesian paradigm will at best consider the statements and descriptions of the ayahuasceros as drug-induced 'hallucinations,' incapable of being scientifically evaluated or verified. From the perspective of a Jamesian radical empiricism however, the phenomenological descriptions of consciousness explorers must be accorded the same reality status as observations through a microscope or a telescope."[1] Are you experienced? an older kid had asked me even in the playground of our elementary school while I idled on a swing. What do you mean, I asked. Hendrix, he explained.

One thing about psychedelic drugs is that they bracket the usually ironclad difference between inside and outside. Under the influence, you can be in a roomful of people and feel you are directly connected to them, a single being distributed through separate bodies. Place and time can be upset, deconstructed as they also are in postmodern prose and the hypertexts of technology.

Reading Doyle on Charles Darwin, drugs, and the evolution of the *noösphere*, this complication of place and time was forcibly brought home to me. In that green house on top of Gibbs Street where I grew up, we had a music room, a little ethnomusicological enclave with banjos and kalimbas and African talking drums. Neatly arranged on its bookshelf were iterations of a journal, the *Psychedelic Review*, whose back issues contained scholarly contributions by the likes of Leary, Metzner, Aldous Huxley, and Alan Watts. I had occasion to gaze on them with curiosity, pulling them out and browsing a bit and taking them en masse as tacit approval of the more-than-street-worthiness of the substances they discussed and were. I did not read them then, but reading Doyle I found that my mother had contributed to one. It must have been there all along. All I knew about her and drugs was that once, when we were teens and her Colombian friend bragged about marijuana, she condescended to smoke a hit in front of me and my teenage brother. I think it was just to impress her friend, and her heart wasn't in it. In the dining room where my stepfather hosted poker games with lawyers, doctors, and academics, she took a little puff, spilling some in the sour cream. "Look," she pointed. "Chives." It seemed to be a joke to cover the fact that she really wasn't interested in getting high on pot, and in fact I'm not sure she ever did. She did tell us later that she had tried LSD under clinical conditions when it was still legal. This was done under the direction of Gene Sagan, an older cousin on my father's side. She said that in the lab setting the men in their white coats looked to her like overgrown babies. She also found she could adjust ambient volume by touching her ears. Gene Sagan, whose wife, Arlene, was a big help to my mother around the time of my brother's birth, later committed suicide by jumping off the Golden Gate Bridge. Not an auspicious start to the scientific investigation of this subject. My mother later visited Leary in his farm in Millbrook, New York. She was not impressed by the sight of him being surrounded by half-naked women, but what really disturbed her was his treatment of his own daughter, who at the time was in a sort of LSD Skinner box Leary had organized so that enclosed subjects could push keys to quantify and record their psychological state. But if she became disillusioned, she was initially excited about the potential for psychoactive drugs to be a phenomenological tool.

As Doyle made me aware, my mother submitted an "open letter" to the *Psychedelic Review.* It reads, in part, "Mescaline is related to adrenaline, a known neurosecretatory hormone; and caffeine is a purine similar to the nitrogenous bases in DNA (the genetic material). If these facts do not, at best, point to physiological mechanisms, they at least attest to the knowability of consciousness, psychosis and mystical experience. The chemistry eloquently testifies to the amenability of man's soul to his own researches."[2]

Writing under her first married name of Lynn Sagan, my mother relates the revelation of the chemical infrastructure of the mysterious substances to the chemical breakthroughs that laid bare the molecular basis of heredity. She argues that, instead of obscuring reality, so-called hallucinogens seemed to reveal it. "The drug attacks defense mechanisms built up carefully to conceal the truth of our direct sensory perceptions. One would *a priori* imagine, however, that a drug which forced us to see the world as it is would be welcomed."[3]

As Doyle points out, "Psychedelics, for Sagan and others, represented a scientific enhancement of human perception akin to the electron microscope, and the early data gathered by this new mode of scientific observation suggested that the separation between the self and the cosmos was an illusion, an artifact of egoic consciousness correctable, reflectable, by a continuously tuning . . . analog consciousness."[4] Although I did not come across her letter (or if I did, in my cursory browse, I did not understand it) in that attractive set of back issues of the *Psychedelic Review* in that Newton Center music room, Doyle's book gave me the chance, transporting me back in space and time, as it were, to my early adolescence.

IN THE THOTH DECK, a version of the Tarot designed by Aleister Crowley, the Magus is depicted with "a naked golden body, smiling, with winged feet standing in front of a large caduceus. In his right hand he holds a style and in his left hand, a papyrus."[5] Crowley, a shadowy figure associated with Satanism and the counterculture introduction of black magick, held that drugs were among the magician's most powerful tools, but he fills the hands of his magician archetype not with pills but the tools to write. Crowley, a mountain climber and expert chess player who disavowed chess when he saw the miserable comportment of the masters, moved on to a nonacademic investigation of the arcane. Although decidedly less on the dark side, my father also related writing to magic:

> What an astonishing thing a book is. It's a flat object made from a tree with flexible parts on which are imprinted lots of funny dark squiggles. But one glance at it and you're inside the mind of another person, maybe somebody dead for thousands of years. Across the millennia, an author is speaking clearly and silently inside your head, directly to you. Writing is perhaps the

> greatest of human inventions, binding together people who never knew each other, citizens of distant epochs. Books break the shackles of time. A book is proof that humans are capable of working magic.[6]

Writing is also a psychedelic. If I say behold the sweet-smelling pink magnolia, or cheap whipped cream, or unseen birds in the spring mist, your mind fills with sensations. You begin to see things that aren't there. It's a powerful drug, all right.

DOYLE'S *Darwin's Pharmacy: Sex, Plants, and the Evolution of the Noösphere* is a book that, like some of the drugs it describes, is hard to put down—and keep down.[7] It comes up, it lingers like a sharp coffee or pungent sprig of rosemary, it catches in the back of the throat and tightens the jaw like some strange hallucinogen that evolved with mammals by intoxicating them, protecting its genetic endowment while attracting the bug-eyed critters just enough to enhance its own distribution across the forest floor and grassy planes.

In mythology both Hermes, the Greek messenger god, and Thoth, the ibis-headed Egyptian god, are said to have brought our ancestors the potent mind-altering substance writers and publishers push. Like its more properly pharmacological psychoactive cousins, writing can produce or break through the screenlike phenomenal layer called *maya* in Hinduism, the layer of ordinary everyday life we take to be real. The Pythagoreans and Platonists and Neoplatonists may have been inspired while on psychedelic drugs to pronounce ordinary reality, after they came down, a charade. Botanically and mycologically altered states may have helped foment the important protoscientific realization that all is not as it seems, that there is more to reality than our waking consensus reality.

The ethnomycologist, and one-time vice president of J. P. Morgan, Gordon Wasson (1898–1986) argued that *Amanita muscaria*, the famous psychedelic and poisonous red toadstool with white spots familiar from pictures of the hookah-smoking caterpillar in *Alice in Wonderland*, was the sacred "god of the gods" (soma) that informed the mystical insights of the Sanskrit Vedanta. Wasson argued that ergoline alkaloids from the ergot fungus, which grow on rye, were taken during the Eleusinian mysteries devoted to the agricultural goddess Demeter or Persephone. Wasson was inspired by his young Russian bride, Valentina Pavlovna Guercken, who surprised him, an Englishman averse to toadstools, by avidly collected edible wild mushrooms. After inciting his interest, the couple made an epic trip to Mexico, collecting psilocybin mushrooms and *Salvia divinorum.* In May 1957 they published "Seeking the Magic Mushroom" in *Life* magazine. The widely read article helped set the stage for the sixties. Today two species of mushroom, *Psilocybe wassonii heim* and *Psilocybe wassonorum guzman,* include Wasson in their names.

I SAY that Doyle's book is hard to keep down because it acts like a drug. By that I mean not only that it makes us hallucinate by conveying the visions that it recounts, which all books with scenic or fictional elements do, but that it partakes of the logic that Jacques Derrida shows at work in Plato when Plato, in his dialogue the *Phaedrus,* writes that Socrates refers to writing as a "drug."[8]

The word Plato writes down to describe the description by Socrates, who never writes, of writing is φάρμακον, pharmakon.

Pharmakon, meaning drug. Much as in modern English, the pharmakon, the root word of pharmacology, can be a poison or a remedy, or neither or both. In ancient Greek it can also mean sacrament, talisman, cosmetic, or perfume. In the *Phaedrus* Socrates criticizes writing as a crutch because it weakens our memory, oratory, and culture by making us dependent on the prosthesis of an external substance. His precipitous judgment was to be turned humorously on its head two thousand years later by the 1960s bumper sticker claiming, "Reality is a crutch for those who can't handle drugs."

As a kind of hallucinatory drug or magic, writing allows us to speak to ghosts, not only nonpresent friends and family members but intellectual influences long since dead. Writing extends, with precise iterability, the ghostly transmissions of language. In what he would eventually call the "stoned ape theory of human evolution," described in the book *Food of the Gods,* Terrence McKenna argued that modern humans evolved some one hundred thousand years ago, potentiated by *Psilocybe cubensis,* the most common form of Wasson's "magic mushrooms."[9] Increases in visual acuity (possibly associated with dilated pupils), aphrodisiac effects (aiding procreation), and, at higher doses, dissolution of ego (enabling socioreligious binding) jump-started human evolution according to McKenna. He further proposes that tripping apes, already grunting and hallucinating, would be perfectly predisposed to start using language, which is, after all, a form of consensual hallucination. Ingestion of hallucinogens would have provided the original natural technology that prepared our ancestors to use sounds and, later, written symbols to communicate, triggering specific images in one another's heads. McKenna's theory is reminiscent of another counterculture motto, "Reality is a shared hallucination."

Whether or not McKenna—or for that matter, the biologist E. O. Wilson, who implicates hallucinogens (and group selection) in the start of human culture[10]—is even partly right, note that writing reworks plants and fungi: Papyrus is the pithy stem of a water plant, and books still depend on fungi-dependent forests whose trees provide the paper on which we write. Recycling and feeding animals, which help propagate and fertilize them, fungi and plants have a long history of evolving enticing compounds that attract our kind even as we help grow them and tend them and spread their seed.

In his book, which triggered these memories and this analysis, Doyle combines scholarly attention to some of the great chroniclers of hallucinogenic and psychedelic experiences—McKenna, Leary, Huxley, and William S. Burroughs—with a recounting of the role ayahuasca played in his own life. It is performative in that, instead of just making an argument, it also seems to aim to create an experience. For Doyle, the terms *hallucinogen* and *psychedelic* are ultimately inadequate. He introduces the term *ecodelic.* As psychedelic comes from roots meaning "mind-expanding," ecodelic comes from roots meaning "home-" (eco is from *oikos,* home) "expanding."

And that is how I felt reading this book—as if our planetary home, to which we owe so much, especially perhaps to plants and fungi that combined to make the ancestral arboreal environment in which humans evolved, were expanding. As if a shroud were being whisked away, scholarship being applied to an absolutely central aspect of human ontology—one that has been unfairly, unsuccessfully, and symptomatically "ritually excluded." Perhaps we have been so blindsided, so preoccupied, so myopic for so long that we first must hallucinate reality before we can see it. What resists persists, and what we cannot digest continues to perplex.

Applying ecological thinking to multicultural use, including Western use, of drugs gives us a new appreciation of and connection to the environment in which we find ourselves. Doyle says he used to be an ontological materialist, taking antidepressants, until he got tenure and, rather triumphantly, because he was going to be paid to do it, traveled to the jungle to take ayahuasca. He had been a great fan of *The Yage Letters,* the correspondence between Burroughs and Allen Ginsberg, and was planning on making a film that went beyond David Cronenberg's movie about Burroughs, *Naked Lunch.* As he researched psychedelics (or ecodelics), he discovered that often people would mention DNA in describing their experiences. This intrigued him, but what really hooked him was when he found out that Francis Crick himself was on a low level of LSD when he discovered DNA's linked spiral staircase structure. Kary Mullis, who won a Nobel Prize in chemistry for modernizing PCR, the polymerase chain reaction that copies DNA in the lab, reported in the September 1994 issue of *California Monthly* that he doubted he would have done so if it were not for LSD. Steve Jobs reportedly said that his use of LSD in the 1960s was one of the two or three most important things he did in his life. Richard Feynman tried it with John Lilly. My father didn't, although I advocated that he should try it and that I would do it with him.

Rather surprisingly, considering their reputation for making people "out of it," ecodelics often reveal hidden depths of our connections to the environment, making us realize we are part of what we observe. Perhaps it is not surprising that Doyle's e-mail moniker is Möbius, the mathematical plane that, given a felicitous twist, has only one side—meaning its inside *is* its outside, quite literally. In "The Wall of Darkness," Arthur C. Clarke describes Trilorne, a world where explorers meet the ultimate snag: an impenetrable black wall extends from the planet's equator indefinitely up. The wall turns out to be a Möbius strip, a surface with only one side, so that all exploration leads to the beginning.

An illustration by the quantum physicist and Albert Einstein colleague John Wheeler (1911–2008) vividly reveals what our truncated, anecodelic objectivity conceals: a giant U with an eye, representing the observing ego, on one stalk. It looks out and sees the serif atop the black pole of the letter's eyeless stalk. The line of sight creates the illusion that the eye is observing something else, rather than an extension of itself.

It is thus as if drugs don't just produce illusions but clear the clutter, removing the artificial feeling of separation between inside and outside, between oneself and others, between life and the environment. In this way they reveal rather than hide, showing us how we are all part of the same thing; rather than simulate or distract, ecodelics can dissimulate and focus: this is perhaps one reason Doyle prefers to call them ecodelics rather than hallucinogens or psychedelics or, worse, psychotropics or psychotomimetics.

I had not met Doyle until he came in October 2011 to Kitchener, Ontario, to be present at the twenty-fifth annual meeting of the Society of Literature Science and Art (SLSA), devoted to "PHARMAKON: That which can both kill and cure." The year before, his plane was delayed to New Orleans, so he missed his place on a panel on plants at the American Anthropology Association. I was happy to give his paper on plant agency, talking about how plants "outsourced" their sexuality to animals, and mentioned how Duke University, the site of one of his recent talks, owed its existence to tobacco plants. It occurred to me that it was a very plantlike thing for him to have missed his plane and be at home in Pennsylvania while I, a kind of ambulatory animal he outsourced, read his words. Not coming allied him to the plants he was representing. It was an Ovidian move.

In Kitchener I gave a reader response to *Darwin's Pharmacy*. I had already been exposed to Derrida's analysis of Socrates's description of writing as a drug. Since Doyle was talking about drugs proper, and the conference was about the pharmakon, I thought it would be fun to investigate the relationship of Darwin's pharmacy to Plato's, to see what might arise from such an alchemical mixture.

As I developed my talk, I was drawn to the subtle changes that occur as we move from speech to handwritten to typed text. In the first part of my response, I read my own handwritten comments:

> This was a great Book—it was hard to put down—keep down—like ayahuasca! Now I shouldn't probably criticize because my objectivity is impaired by contact high—plus drug interaction problem—I'm also a mycelial thinker!

But really, Doyle's book is an important contribution to an important, marginalized discourse. It's really about ecology and psychedelia. It's serious and playful at the same time.

What *are* these drugs that you talk about? What are they *saying*, *to us* and *through* us?

Now this is how I talk, not how I write.

Writing, remember, is a drug—according to Derrida, according to Plato, and maybe Freud—and according to *me*. And that's a good thing—maybe!

We eat plants to live and fungi recycle our bodies, preferably after we're dead. But sometimes we don't digest them; we throw them up. They don't *sit well*; they intoxicate, entheogenate, neologize us.

McKenna, hyperbolically—and one of the great things I loved about Doyle's book is that he's read so much McKenna and so much Philip K. Dick, people I'm really interested in, that it's a great way to get a distilled dose. It is kind of like whiskey or cognac. I admit I'm a card-carrying literary heterotroph, by which I mean I like to go to the top of the literary food chain and feed on it. That's good and bad because the mistakes are multiplied and magnified; at the same time, you don't have to do as much research, which is very important to some people—McKenna, hyperbolically, but with clarity brought on by pupil-dilating psilocybin, says language itself is a shared hallucination jump-started by monkeys on mushrooms. You know, the kind that grow out of bullshit, literally.

And when we don't digest them, they can entheogenize us, take us on a trip. Doyle shows us where—right here! But in his book he is on drugs—on the ur-drug of writing.

I read it on a Kindle so I'm safe, but it's still writing and I still shared Doyle's trip—going everywhere and nowhere. Now, according to Derrida's theory of a general writing, we've always been on drugs. And note that writing in its original form is on paper, undigested or technologically partly digested vegetable matter outside our bodies. The very cellulosic and lignin strength that allowed algae to stand straight, cyanobacteria to erect monuments that allowed their fragile, wet bodies to survive the dry earth, like pharaohs reborn in a solid realm, allows us to preserve and repeat traces and trails.

Biblion in Greek "names the internal bark of the papyrus and thus of . . . paper" and is "like the Latin word *liber* which first designated the living part of the book before it meant 'book.' But *biblion* can also, by metonymy,

> mean any writing support, tablets for instance"[11]—and this was written before the iPad, which I don't have. Like I say, I have a Kindle. I'm a Luddite, a late adopter.
>
> But we are all using drugs here, Gaia-given mind-altering ecodelics. And if the government were consistent, as Doyle suggests, it should put itself in jail for possession of a dangerous mind-altering substance.
>
> Despite his playfulness, Doyle's message is, "Let's *take drugs* seriously, they have an ecological message."
>
> So I also have some typewritten comments, but this is handwriting so I'm considering this like a gateway drug to typewritten drugs. I don't know if we want me stop there, or if you want me to talk about thermodynamics, or read my—it's funny because when you talk, and when you write with your hand, and when you write with a typewriter it seems to actually affect your communication, right? I mean, we don't talk like we write; it's an amazing thing.

Encouraged to continue, I read my typed comments, commenting that I'd probably read faster, in part because I wouldn't have to try to figure out my handwriting.

TOWARD AN ANAXAGORAN ECODELIC PHARMOTOPIA

If writing is a drug, as Derrida emphasizes—and bear with me, this may be as convoluted as your brain—if writing is a drug, as Derrida emphasizes, perhaps I should say *pushes* on us in "Plato's Pharmacy" in his explication of Socrates's take on manuscripture's dual form as remedy and cure, mnemonic and crutch, Doyle's *Darwin's Pharmacy* is a trip. It is a rhetorical explosion. The Christian evolutionist geneticist J. B. S. Haldane said infamously that the world is not only queerer than we suppose, it is queerer than we can suppose. But Doyle helps us, with some help from *ayahuasca* and his anthrobotanical, biospheric, and historical friends, to imagine this unimaginable. This is literally trippy, less in the 1960s cliché sense of *far-out* than in, to use a phrase from another sixties figure, Alan Watts, far-*in*.

It is a trip but, as the American transcendentalist Ralph Waldo Emerson points out, "There is no place to go." Doyle's electropaper hybrid text is neobombastic—in a good way. Remember Bombastus was the name—one of them—of Paracelsus (born Philippus Aureolus Theophrastus Bombastus von Hohenheim), the alchemist physician par excellence who emphasized the notion that "the poison is the dose."

In trying to place the perplexing neobombastic difficulty and transpharmaceutical euphony of Doyle's text, I am reminded of his own elucidation of the notion of a hyperbolic plane, a geometric surface that is the opposite of a sphere, in which every curve, instead of folding into itself, moves away from itself, infinitely.

This impossible figure that is the opposite of the sphere seems the perfect complementary trope, or perhaps antitrope, to match or complement the very striking and real phenomenology of noös, of consciousness, of spirit versus matter, the mind we live versus the flesh we have: what could be a more perfect complement to the absurd physical reality of all of us eensy-weensy beings, us Lilliputian, Horton-hears-a-who, antlike nanoprimates sniffing technological pheromones on the surface of this tiny blue sphere in the depths of unimaginably vast and indeed perhaps infinite space, than the equally, and oppositely, grandiose figure of our spirit—our spirit on the drugs of writing—as a hyperbolic plane?

One fills the other, complements it, suggesting that, even if Emerson was on to something, that there is no place to go, that we are where we want to be, running beautifully and hyperbolically in place, on the surface of a sphere in the depths of possibly infinite space. It is a heady feeling. Doyle brings us there, and I am grateful to him for upping the dose.

For Derrida, following Socrates but diverging from Heidegger, who lauds Socrates as the most perfect philosopher because he doesn't write, writing is a pharmakon, a drug, a supplement that obeys the double logic of aid and handicap, window and screen, welcome addition and threatening replacement. The Pharmakeus is also a magician or medicine man, cosmetics, and the Socratic figure of he who is ritually excluded from the socius—unable to be digested and assimilated by the body politic in a way, formally similar perhaps, to that of drugs like *Amanita* or ayahuasca in the body.

These pharmaceutical and philosophical glitches and intrusions that are so hard to swallow personally and socially get us off; they send us on our way. Derrida talks of a general writing, an arche-writing,[12] and writing in retreat, retracement, withdrawal. He, like Doyle, might take exception to the Heideggerian hand, which gathers and makes humans nonpareil and metabiological, somehow safeguarding our superiority over our biosphere-mates.

But, as the sound bite spirit master Eckhart Tolle reminds us, "I have lived with several Zen masters, and they have all been cats." It is too early in our postreligious hangover to recapture human superiority under the umbrella of a Heideggerian ontotheology or the undeconstructed scientific myth of humans as "the highest species." In philosophy as in science, where there is a movement afoot to name the present age the Anthropocene after its human namers, our hubris is showing. Consider the source and Doyle does, showing we are anthropobotanical and anthropofungal nodes in the noösphere. We do not know what we are. Not knowing is the great crossroads of science, religion, and philosophy, the creamy part of the Venn diagram that will sustain us as we lurch forward into the cosmic unknown with something more than hope but less than speciescidal self-confidence.

For Samuel Butler, technology itself was also a kind of drug, obeying Derrida's double logic of writing as supplement, tough-to-swallow psychotropic, ecodelic, entheogen.

Arguing with himself in the pages of the New Zealand papers under various pseudonyms, the wily Butler described technology at once as an extension of the limbs—and now, with computers, we would say mental faculties—of the human race and as our biggest threat, "stealing across the face of the globe," as he put it, like a new species out to replace us. From writing, which aids but destroys the memory by storing memories outside the head, to the electronic calculator that improves our calculative abilities in the same way, to television robbing kids of books, this double pharmakon structure keeps coming at us. It looks, feels, and smells evolutionary. And indeed it may well be.

But the very first—originary, if you will—example of this pharmakon structure is Doyle's: writing, on paper, is a plant product, one we don't digest, outside our bodies. This capacity for informational excess outside our bodies from which there is, as Derrida argues, no exit, is vegetal. It founds us, literally, as historical beings—preserving a written record. The plant world is also the world from which we primates come, from which we emerge—into houses of wood and stacks of paper, clothes made of textiles—into civilization concordant with agriculture and history. A species made possible, brought to you, in a quite real sense, by plants—and other beings. As nothing is wed to being, as fish swim in water without seeing what is so close to them, as we breathe the plant-oxygenated air, so the paper of writing acts as fantastic hallucination-realizing pharmakon. It is an underappreciated plant gift. Doyle helps us realize it and shows us that, nightmares of the paperless office notwithstanding, plants are not finished giving. Like Marlon Brando in *The Godfather,* they are making us offers we will be hard put to refuse.

Here's an idea: The world is a perplexing mix of the finite and the infinite. It is an infinite mixture. But it can be unwound, disentangled with noös, mind—the noös of the noösphere. This is not my idea, but that of the pre-Socratic Anaxagoras. It was preserved by the medieval church in the doctrine of indwelling of the three persons of the Trinity. I learned from Doyle that Anaxagoras agrees with Empedocles that plants move according to their own desire. His idea of perichoresis—that is, Anaxagoras's idea of the universe as infinite mixture that can be unpacked by noös, the mind of the noösphere—suggests our minds are on to something. Perhaps they will ultimately be able to retrieve or re-create the primordial state, getting us back to the Garden along a Möbius Trip. When humans think, Gaia laughs: our thinking and technology are all part of something bigger, older, deeper. At the same time the noösphere is manifesting this something.

In its secular form this infinite mixture unstirred by mind and the work of the ecodelic noösphere is accompanied by clear evolutionary vectors: increase in aggregate numbers of organisms, their types, connections, aggregate information-processing abilities, sentience, ability to perceive new gradients from which to extract energy to maintain and expand themselves in complete accord with the dissipative tendencies of thermodynamics' second law—the centerpiece of that body of physics theory that Einstein said was most likely to resist future change. Here we may find Doyle's peacock feather and Shiva hallucinations, the cross-cultural, biospheric truth of them. It is a difficult Vernadskian, Bataillean truth in which the mind that perceives it is revealed to be not central but a means along with sex and death to expand the possibilities of the biosphere, to explore and develop new regions of evolvability, to find new ways to spread genes, recombine genomes, arrange groups and tribes and multispecies assemblages to form working solutions to the cosmic task of ecologically sustainable gradient reduction. The second law tugs organisms to find ways to work together to stably reduce gradients, dissipating the energy that sustains them. The "sustainable" part is crucial but often gets lost in the work of thermodynamic theorizers and those who would critique them as making an unreconstructed unthinking and politically dangerous contribution to neoliberalism, as objectionable in its way as is social Darwinism and neo-Darwinism's caricature of Darwinism.

What these critics and theorists miss is that life's accomplishment of an implicitly destructive thermodynamic mandate—to seek out energetic gradients and lay waste to them, accelerating the inescapable work of the second law through a kind of daimonic biotic intelligence—is always ever compromised, threatened, by its own success. This is the origin of prudence and long-range planning, of moderation and likely also of the natural means of population control we call aging and programmed cell death, which are now known to be under multiple forms of genomic control.

Don't worry; we are starting to come down now, only a few pages left. Indeed, the Shiva factor that Schumpeter calls creative destruction and Bataille calls expenditure without recompense is modulated by an ecological intelligence that is less fully developed in humans than in the ecosystems your species is destroying to make way for hamburgers. Human techno-intelligence is phenomenal, plastic, amoral, capitalist, and dangerously Promethean. We are overly impressed with it; it may be a flash in the evolutionary pan—like those bright flames magicians produce with chemically treated paper, that leave no ash but make a nice segue to something else, like a dove or a flower. Vernadsky pointed out before World War II—he was struck by the movement of munitions in World War I—that the number and rate of transport of elements in the Periodic Table involved in life's process in and as Earth's surface augmented over billions of years of time. With high-energy physics, fleeting isotopes never seen in our solar system before have been produced.

Today the situation is better-worse: modern atom-smashers such as the Large Hadron Collider produce energy levels seen elsewhere in the universe only in black holes and supernova explosions. Indeed, Otto Rössler (who brought Anaxagoran perichoresis to my attention) lobbied to stop CERN because of his calculation of a one in six chance of its forming a mini black hole that would inadvertently destroy the Earth. The romantic irony of this technical apocalypse makes it now my favorite apocalypse after J. Marvin Herndon's nuclear geomagnetism failure scenario, and that only because the Herndon scenario features a Zen-like breakdown of global telecommunications accompanied by beautiful planet-wide aurora borealis effects resulting from the depletion of uranium that, undergoing fission in Earth's core unbeknown to current science, is the actual source of our protective magnetic field. Perhaps both scenarios are preferable to the death by a thousand cuts theoretically being engineered by a globalist sociopathocracy scarcely to be preferred to mob rule.

On the positive side, nuclear lasers in space, it has been published, could be used to generate black holes at a safe distance. If, as some think, new cosmoi, cosmoses, arise on the other side of these black holes, then we confront a strange possible conflation of physics and metaphysics: is this universe the result of previous universes that have multiplied via the evolution of black hole–forming technonoetic polities?

Are we godlets in the making? Maybe, but right now it don't look so good. Our chattering monkey minds and taste for technical sweets are not in accord with Team Gaia, the long-evolved, long-lasting, gradient-reducing virtuosities pioneered by ecosystems, including versions of the same forest ecosystems that fostered our own "recent" evolution.

As Schneider showed and Doyle reminds us, human gradient-reducing abilities pale in comparison to those of plants, especially those of diverse communities such as found in Amazonia and Borneo.[13] This is not theory but thermal satellite measurement showing cloud cover temperatures over rain forests in the height of summer to be cooler than over Siberia in midwinter. Cut down a forest and it will get hotter. These plant-based biodiverse enclaves are, moreover, more stable in principle—more like adult bodies or healthy organisms than pioneer species or pathogens sweeping through a body or ecosystem. They are multigenomic, calm and biodiverse, not monoculturists metastasizing the countryside. Plants are our great planetary gradient reducers—and, yes, they perceive the solar gradient, turning their leaves and flowers to follow the sun. Indeed if we accept the evidence for anthropogenic global warming, it is clear that humans, despite our minds, have depleted global gradient-reducing function by increasing temperatures near the surface. Human technics are amazing but literally globally dysfunctional.

As Doyle also points out, there is a question as to the role of what Darwin called sexual selection in all this. Darwin was criticized by Butler and others for being overly mechanical, capitalistic, and simplistic. It seems likely that in trying to carve out a naturalistic view of nature in distinction to the superstitious teleologies and anthropomorphisms of Christian religion, Darwin accentuated the elements of evolutionary theory that worked themselves, the elements most akin in principle and spirit to the mechanisms and mathematics of a Newton or Descartes. There is thus, in general, little role in Darwin's careful speculations for subjectivity or the effect of individual conduct on the evolutionary trajectory. Aside from humans, whose behaviors and ideas clearly influenced social organization, the one place Darwin felt compelled to grant a shaping influence was in mate selection, especially by females.

At the end of the Kitchener session, responding to my response, Doyle asked about the thermodynamics of beauty: "In neo-Darwinism," he said, "natural selection and sexual selection are conflated. But then you'll see a fan of peacock flowers and [it forces you to ask what is the source of this] exorbitance [in] squandering a gradient. [The peacocks seem to overlook] what they do. What they do is grab attention. You see a sort of tug toward the ability to grab attention and therefore [is it not the case that] beauty is in this evolutionary biological sense an ongoing race to be able to grab the attention in some fashion? Flowers do it, peacocks do it, humans being singing songs do it. Don't we see, as Miller shows,[14] what is a linkage to the importance of human consciousness and that this explains in part the function of human consciousness [which] can create infinitely unfathomably beautiful things that constantly elude our total grasp and force us onward to squander more amounts of information and hopefully sustainably? The peacock does it sustainably, so I wonder what you think about that informatic beauty part."

I answered by comparing the difference between sexual selection and thermodynamics, on the one hand, and natural selection proper, to the difference in philosophies between Arthur Schopenhauer and Friedrich Nietzsche. Schopenhauer emphasized the will to live preservation, whereas Nietzsche expanded this idea to the will to power (the name of his collected notebooks). This philosophical dyad parallels the difference ecologically between sustainability and the Bataillean realm of excess that one experiences as pure thermodynamic gradient reduction. It seems to me that both these registers are working and that in evolution there is a sort of experimental realm that is tied to thermodynamics and goes beyond mere survival. If you look at it scientifically, it is clear that we have to stay alive. Otherwise we're not going to have offspring. This is the Schopenhauerian will to live. But at the same time there's a thermodynamic drive and we are only alive, life only exists in the first place, because of this. That is to say, life forms a more active region, a metabolic zone of energy flow that perpetuates the firelike spread of energy formally described by the second law. If we can find ways to tap into gradients, organisms—all organisms, not just us and our external technology—will expand as quickly as possible. This is the natural tendency for unchecked growth and exponential reproduction that natural selection curtails. However, it is not necessarily a stable regime. The difference between the Schopenhauerian will to live and the Nietzschean will to power is the difference between a sustained flame and a raging fire.

Sexual selection is more like the out-of-control fire. It has been described as a runaway phenomenon where you can select for features that don't have survival value. A peacock is more likely to be attacked by predators because he is more noticeable, not only to females but to predators. That runaway can be a dead end. But these regimes of rapid growth favored by the second law over and above survival can be aesthetic, artistic, crucibles of experiment and innovation from which new stable regimes emerge. When Doyle did the unthinkable, calling the War on Drugs into question, an act for which one might be denied tenure, and looked through a scholar's eyes at psychedelic discourse, he was surprised to see so many references to molecular biology, specifically to DNA. It seemed like it might be more than a trope, the metaphor of the age; in both molecular biology and psychedelic experience information was less about discrete meaning than a recipe or algorithm for change.

They seemed to speak the same language, a language of information that helically revolves around itself, that is convoluted and more active. I agree that this parallel may be more than a coincidence. I think the realm of play and informational excess might be seen as a cutting edge of new regimes of evolvability such that short-term expenditure of energy and excess has no survival value per se, and is closer to the runaway phenomena of sexual selection than the more staid realm of evolutionarily stable strategies. Nietzsche said, "Existence is only ever justified as an aesthetic phenomenon," and it is these aesthetics beyond survival that one notices while tripping.

A nice example, *A Short History of America,* a twelve-part comic panel by the underground comic master Robert Crumb, illustrates the role of energy flow and perception in an experimental regime of excess beyond the realm of survival, of mere preservation or even evolution as commonly understood.

The panel begins with a bucolic countryside, trees and birds, unmarred by humanity or its artifacts. As our eyes scan across sinistrodextrally, we see this same hillside cut by a train, then a road with a horse-drawn carriage, and on and on to telephone wires, paved streets, houses, and cars. In the end, there is a city with a Texaco station, billboards, tenements, streetlights, and a pink pimpmobile. Here we see a realm not of natural selection per se—the drunken oil economy has not yet been selected against—but of expanded energy use and networking based on gradient perception, the perception and exploitation of available energy gradients.

The immediate realm of attention in the biosphere, "subjective" and dispersed among myriad organisms, is more directly connected to gradients in the environment than to long-term evolutionary effects, although of course it does have evolutionary consequences, as we see in sexual selection where mate choice reinforces attractive features by reproducing them. If you posit awareness all down the line, not just in human beings, then you have this constant awareness going on that is connected with the use of energy, which is theoretically what the awareness is for. Ecological function demands I find ways to get the materials I need for my body to go on, and that would be like Schopenhauer's will to live; but also in play is the question of how can I expand my realm, how can I use more available energy. But this latter can be a dangerous thing, too, because expansion, like a fire, can lead to burning out; it can destroy the system that's doing it. So there's a play of forces. In the work of Jeffrey Wicken on information and thermodynamics, he talks about the two very important drives, not just to survive, which is in the Schopenhauerian register, but also to use energy.[15] And so sustainability is a sort of a version of the Schopenhauerian will to live, preservation, the survival with which we're familiar in Darwinian terms. It dramatically contrasts with sexual selection, of excess. The Nietzschean–Bataillean realm of aesthetics, energetics, and excess does not have a direct survival component. In fact, sexual selection seems more directly connected to what is informing it, scientifically underlying survival itself, which is use of the gradient. What you love can also kill you, what feeds you can destroy you, and that's a sort of a condensed version of the phrasing of the paradox of our relationship to external growth, which is governed by the second law of thermodynamics, and which evolutionary stability aids only because it prolongs a process that would disappear without it.

Thus one of my slight objections or criticisms of Doyle's book is that he conflates information—and this isn't just him, this has been done for a long time—with energy. The equations in thermodynamics and information theory are almost identical. Obviously we data process as living beings, and we also use energy. They're not the same thing though. There's been a conflation since John von Neumann told Claude Shannon, "Oh just use the term entropy, nobody knows what it means anyway." I have a friend named Frank Lambert who has the most visited sites on entropy and thermodynamics on the web. He is over ninety years old, a professor emeritus at Occidental College who worked at the Getty Museum in California. As a chemist, he has a very nice simplification of the second law that applies to both the open and closed or isolated, thermally sealed systems.

The basic sense of the second law is very phenomenological. It's not mysterious. You don't need abstract nouns or shibboleths or a PhD to understand it. You don't even need to reify it with terms like *entropy*. Entropy is a measure of the tendency of energy to spread. It's like an odometer. Entropy doesn't run anything. It's a measure of the spreading of energy.

Another basic popular misunderstanding that even scientist types like Daniel Dennett make is that there's no direct relationship between disorder and the spread of energy. As complex beings that have energy flowing through us, we need to be and are very complex. We stay organized, and that organization is directly connected to our use of energy and our dissipation of energy. So there's a whole bunch of unnecessary confusion about the energetic essence of life as a complex system that is phenomenologically attached, sensuously involved and perceiving of its surroundings which it helps bring, slowly or rapidly, closer to equilibrium. This is because of the science of thermodynamics, which is kind of like economics: it's a sort of dismal science, but intellectuals in the humanities love to appropriate thermodynamic terms because it's fun, it's sexy, and, like von Neumann stated, nobody knows what it is anyway. But this academic fetishizing misses an opportunity for a real communication about the basic nature of our reality as complex energy beings whose perception is related to our metabolism and gradient reduction. And I think that it is this direct phenomenological connection to our status as gradient-reducing systems—which becomes highlighted and magnified during the would-be "hallucinatory" psychedelic experience—that is so exciting to both me and Doyle. The thermodynamic worldview revalidates and contextualizes our real-time connection to an energetic living world. The more-than-human, more-than-living flows that underscore metabolism and survival are operative in the real world, and the thermodynamic description of life intuitively rings true. There are energy gradients; they're tapped into by living beings; we help them. Not only does life not violate the second law of thermodynamics, but we actively perpetuate the spread of energy. Even nonliving systems like Bénard cells, Belousov–Zhabotinskii reactions, and storm systems have a fledgling identity, a temporary stability. We recognize a kinship to them. They are fellow complex materially cycling systems. The helically coiling DNA molecule itself looks like something out of an acid trip. But these vortices and complex cycling systems do not need to be tripping, nor do you need to be tripping to perceive them. They are part of a creative destructive realm proper to the deconstruction of gradients and the pleasures that can accompany that. If seeing vortices and spirals and iridescent false eyes of peacock plumage seems to go beyond mere survival, it is because it does. It tap into a more-than-hallucinatory moment of active energy degradation, the realm of which comes before either life or death. Temporarily stable complex systems more actively degrade ambient gradients, but to do so they've found ways to maintain their perimeters within the flux they enable. Organized, they are cyclically churning matter. Watts understood nonequilibrium thermodynamics intuitively, linking ecology and psychedelic experience as he described us as skin-encapsulated egos that have forgotten our connection to the whole, our kinship to the whirlpool and the flame, to the reality that we are glorified tubes taking in organized matter at one end and putting out degraded matter at the other. Our bodies are less temporary than a whirlpool; more long-lasting than a match zoomed in on in a David Lynch movie, but still, we are essentially processes, not things.

In a profound way the thermodynamic worldview is both a teleological and a nonteleological view: Yes, we're doing something, we're going somewhere, and you can look at this over the history of evolution and see phenomena like Crumb describes—an energetic spread of energy-using life-forms, including human beings and our technology. Like I mentioned, there are more and more organisms, there are more species, they inhabit a greater extent of the environment, there are more elements in the Periodic Table that are involved in the biosphere, and there are more forms of interconnection. It looks like the Internet is a form of planetary growth, humanity is part of it, that life is becoming more interconnected and symbiotic. So there's these vectors, and it's not true to say there's no direction to evolution, as people as superficially opposed as Stephen Jay Gould and Richard Dawkins have said. They both basically agree that evolution is random. But this is a mistake, both tactically and empirically. There are creationists who say, no, no, it's not random, there's something's going on, it's going toward us. But that, too, is obviously wrong. They have a point, but the large point is teleology does not end with us. We are riding the crest of a wave that looks like it's going straight past us into technology. As Freeman Dyson has said, if life gets a toehold in space, it may kick off its human shoes.

So I think it's exciting, both scientifically and phenomenologically, both for the straight and the stoned, that there's this thermodynamic vector. And it's connected in a way, I think, to the difference in philosophical emphasis between Nietzsche's will to power and Schopenhauer, his teacher's will to live. Nietzsche came afterward, but his will to power is more originary in that energy is more basic and sets the context for life.

So I agree with almost everything Doyle has said and especially the beautiful tenor and complexity of his seriously fun discourse, which repeats as Anaxagoran traces on plant product—paper—elements of the functional and transhuman and, as Doyle himself incarnates, suprahuman biodiversity and complexity of the Amazon itself. It is not just a pharmaceutical hallucination, this trip we're on. The rain forest has a lot to teach us and seems to have found a most eloquent and hardworking and good-spirited spokesman in Doyle by which to do so. We should listen—and read.

AFTERWARD DOYLE COMMENTED that we're overly impressed with our technology, that ethnobotany is in a shamanic war against pharmacy companies and the question is which vision of human being is going to win, and that the answer is there's no contest, because the pharmacy companies don't have a vision of the human being; they only have the next prescription. In Kitchener he argued, correctly I think, that each of us as a human being is a local instantiation of something much larger, that if more of us started thinking of and experiencing ourselves on this larger scale, we might become more sustainable as a natural energy-using form, part of the biosphere and developing noösphere.

If we did an evolutionary–ecological interpretation of this stage of capitalism, looking at these ridiculous pharmaceutical commercials and companies that are pushing these drugs and trying to patent life and chemicals that are endogenous to our brains, one might argue that, well, this capitalist moment in evolutionary history actually worked because it allowed us to tap into these gradients and get all these tools going, some of which might be used for life as a whole. But still, we should not be too quick to assume that they—the global nexus of other organisms in their concerted activities—are going to renew our lease. Now, if we step outside our human box—something enabled by taking rain forest drugs seriously—and look at the situation from a more expansive framework, we might be tempted to say to our species, relax and stop burning things down. You're going to have to either get with the program or take a hike now that we have the technology. From a biospheric point of view, human beings may have served their purpose as an incubator of technology. We've had it nice; our overly self-focused growth was nicely enabled by a capitalist model, but this solely human-focused growth doesn't necessarily have any evolutionary legs. We are drunk on biopower, but we are not necessarily holding a winning ecological hand. That would be a possible interpretation opened up by the more expansive, culturally inclusive, ecologically responsible view of psychedelics brought on by Doyle.

CONCLUSION

FLOATING INTO SPINOZA'S OCEAN

G. EVELYN HUTCHINSON, considered the single most important author to understand the fundaments of modern ecology, emphasized that a scientific theory's primary value was not its usefulness but its ability to produce a form of enlightenment, similar to a great work of art.[1] And while Friedrich Schelling, the German idealist who tried to respond to Baruch Spinoza's ideas on freedom, reportedly said there was no use in criticizing a philosopher for being incomprehensible, I think it's better to follow Richard Feynman's advice—that if you understand an idea you should be able to explain its gist to your grandmother. The idea of life's basic purpose is not hard to grok. Life is a special case of the general phenomenon described by the second law of thermodynamics.[2]

Located between the sun and space, life taps into an electromagnetic gradient defined by a difference between short-wave and long-wave packets of quanta. Like other complex systems, it finds ways to spread energy, producing entropy as heat, as it maintains and grows. You pull over to the side of the road to use a rest stop. A streamer of air finds its way out through an electric outlet into a cooler room. This is purposeful behavior. There is nothing mystical or specifically brain-based about it, although brains turn out to be quite good at helping find gradients, as the present melting of the ice caps shows. Demystified, purposeful behavior simply describes actions that take place in order for future states of affairs to occur. And the mother of all these future states, the main one, is equilibrium.

That processes have a natural purpose, or telos, does not preclude them from being polyteleological. Even trees have other purposes than spreading energy, for example, producing fruit to attract animals or even chemical messages to warn other trees of an insect infestation. Purpose need not and should not be conflated with consciousness. This applies even to our own behaviors, as Pascal suggested when he said the heart has reasons of which the head knows not.

Understanding the universe not as completely lifeless and soulless (so to speak) but as a telic medium in which we are deeply complicit is a breakthrough for understanding ourselves and our universe, and it provides a springboard for a richer view of who, what, and why we are. It is deeply consonant, I think, with the *philosophia perennis*—the perennial philosophy—that we are at one with the cosmos.[3]

It also leads to valid scientific research programs. As the Spanish physicist Carlos de Castro Carranza points out, our understanding of ourselves and others as organisms prompts us to ask productive "why" questions. Embracing organicism, a world of real living things as opposed to bodies that are only validly explained by mechanical metaphors, focuses our attention on function and the future, on the register of why rather than how. J. Scott Turner, for example, wonders why termite mounds control their temperature.[4] The termite mound and the beehive and the ant hill and the human city are in a sense not just the "house" but a "technological" extension, part of the body of a genuine superorganism. The evolution of detrivores and homeiotherms increases life's gradient-reducing means. Teeth, as Aristotle said, are not just random but *for* something. Nor, perhaps, is Aristotle's example exactly random, for teeth are the part of our bodies that are perhaps most related to entropy production. They get at gradients, increasing life's ability to lay waste to concentrated energy reserves. They "grow by necessity, the front ones sharp, adapted for dividing, and the grinders flat, and serviceable for masticating the food. . . . Wheresoever, therefore, all things together (that is all the parts of one whole) happened like as if they were made for the sake of something, these were preserved . . . and whatsoever things were not thus constituted, perished and still perish."

Looking at the whole leads to a different understanding than focusing on the parts, both at the biospheric and at the cosmic scale. This is important in part because it leads to the possibility for a new, rational confluence between science and spirituality. Alfred North Whitehead was of the opinion that more than on any other single factor, humanity depends on the way science and religion interact. Gregory Bateson in his notebook ruminations on the great eye of Leonardo da Vinci, whose own notebook drawings gravitate as if magnetically to the heart of the nontrivial, wonders about the Renaissance master's unerring prescience and focuses particularly on the Italian's investigation of the wave. This leads Bateson to exult in the figure of the wave, this "hump" across the water that maintains "its" shape as "it" travels. He asks if we, if you and I, if the pronoun standing for discrete human identity, is itself not just such a wave. What is this "it," this "hump" that consists, as do each of us, of different particles as it develops and eventually crashes, returning its droplets to the wide and frothy ocean? The pronoun is shorthand. The waveform persists. It is not unlike a glorified, thinking version of a flame or whirlpool. You don't have to believe in anything more mystical than the naturalistic reincarnation of cells in variant forms including children—the curling of Batesonian waves across a cyclical sea of hydrogen, carbon, nitrogen, oxygen, and other atoms—to believe in life's prosaic purpose. Nor does such a scrappy creed disqualify you from the right to be an Ophite, a Gnostic, or a Presbyterian.

Carranza argues that the organicist view brings up distinct but equally scientifically valid viewpoints that can't be whittled down to mechanism. This is probably because a view to the whole makes us focus on function in ways that we don't when we focus on the part. For example, a clearing in the forest is better seen as a kind of scar that is healed by the recolonization of trees in an ecosystem than as a naturally selective competition—just as it would be silly to try to explain a wound healing on the hand in terms of natural selection among blood cells.

Our organicist status connects us as teleological beings to a teleological universe, at least at the level of the part. At the level of the whole the universe may be purposeless and eternal, infinite and multidimensional, as Spinoza suggested. But even at the great levels of exploding supernovae and black holes, teleological maneuvers may be in evidence.

James Gardner vigorously argues (he used to be a lawyer) that we are technologically on course to create black holes, and therefore may not be an existential fluke but part of vast cycles of universe creation. Scientific papers speculate on creating black holes in orbit using nuclear lasers, and I already mentioned that Otto Rössler fears they may be inadvertently created by CERN. If intelligent life carbon-based forms like us can produce black holes, not right away but in a geological moment, then maybe that is why we are here: to produce *again* (though we don't remember the last time) new universes. Elaborating on the physicist Lee Smolin's idea that new universes with distinct laws and physical constants, some more conducive in turn to further black hole production, Gardner adds the wrinkle that life could be part of this process, that it is not doing nothing but integrated into the reproductive workings of a "mother" universe doing its cosmic shebang.

Similarly Rick Ryals, an autodidact whose father was an electric engineer, argues that "we are here to do work *efficiently,* and at the highest level of our technological capability, we create particles that directly affect the symmetry of the universe. That's something that very few sources can accomplish, and it is directly connected to the structuring of the universe, which is why there is an implication for a bio-oriented cosmological principle."[5] Particle accelerators create energy levels not found elsewhere in the universe except in black holes and supernovae explosions. Based on his study of certain basic ideas to do with the cosmological constant and certain anomalies (e.g., quantum field theory's prediction that the vacuum energy density of our universe should be about 120 orders of magnitude greater than is observed), Ryals rejects multiverses and tables the idea of a

> bio-oriented cosmological principle [that] extends well beyond Earth to include similarly balanced planets like ours, elsewhere in the habitable zones of the universe. This is a falsifiable prediction that strictly constrains the physics to a limited slice of space and time, and it explains the Fermi Paradox as well, since all life in the universe will have appeared at approximately the same time in its history, so we will all be similarly developed, technologically. That also means we will all be performing the same function in the thermodynamic process, and that means that we are a collective force to be reckoned with when it comes to our ability to directly influence the

> evolution and symmetry of the universe, itself. . . . The most radical scientific implication of a direct connection between carbon based life and the mechanism that defines the structure of the universe, itself, is the evolutionary mechanism, which, when taken inversely, says that the future universe doesn't eventually die from heat death, and it doesn't "crunch," or anything else that has been proposed, rather, it "evolves" toward absolute symmetry, via periodic big bangs, to higher orders of entropic efficiency so that energy truly is conserved in the never ending effort toward absolutely balanced symmetry between the vacuum and matter. The next universe will be just a little bit flatter, (closer to zero net energy), and will, therefore, be that much better at disseminating energy, over a "that-much" longer of a duration, allowing energy to be disseminated "that-much" more uniformly before the next big bang starts the process over again. This can be understood to be a "downhill" process, because it will require less energy to create the next universe if the new universe is structured to disseminate energy more efficiently because it won't have to work as hard to accomplish this. The physics for this is very simply *real* particle pair creation from vacuum energy, like you get with Hawking Radiation, which causes the vacuum to become more rarefied and "stretched thin" until this process compromises the integrity of the forces that bind the universe.

Functional questions lead us to think of how parts fit into wholes, wholes that in nonequilibrium thermodynamics are not things but themselves processes. What is the role of a cell in the body, of bodies in populations, of people in societies, of species in ecosystems? Understanding ourselves as exponents of a natural, rather than special teleology, is, I would argue, a scientific satori that gets to the root of the connectedness of all things. It brings together Darwin and Spinoza, Gaia and microbial ecology, phenomenology and materialism.

In *Rocks of Ages,* Stephen Jay Gould writes that "the conflict between science and religion exists only in people's minds, not in the logic or proper utility of these entirely different, and equally vital subjects. . . . A blessedly simple and entirely conventional resolution emerges. . . . Science tries to document the factual character of the natural world, and to develop theories that coordinate and explain these facts. Religion, on the other hand, operates in the equally important, but utterly different, realm of human purposes, meanings, and values."[6]

Rather than consider spirituality and science as "non-overlapping magisteria," however, I see them integrally linked. Indeed, future diagnostic manuals might diagnose as pathological those who do not integrate the logical and the numinous, as suffering from a malady, not a French sickness of the soul, malaise or ennui or anomie, but a chaotic theopathy, say, some deficit of Whiteheadian cosmonoia.

HOW MANY ANTS I killed yesterday. They were gathered at the ground zero of an organic honey gradient, milling about the sweet outdoor sculpture of a seemingly empty spoon. In a fit of automaticity, I dispatched them from the linoleum countertop with the pullout sprayer, sweeping them to the slippery curvature of the metal sink. They collected in the stainless steel strainer, a few running for their lives up the steep sides. One or two I watched escape, their tiny, agile limbs vigorous as they abandoned their brethren. Today I saved a bee. Caught, it bounced between the panes of a lifted window, unable to solve the simple maze. Deserving to be stung for my sins of the day before, my hand cupped its furry thorax, its forewings aflutter as I solved the maze for it, opening the bars of my fingers and giving it, after a brief hiatus in midair, a tap on the behind to send it flowerward into the spring.

How evil and good of me. But what guided the hand I watched withdraw that stretchable faucet? What thought process or impulse drove me to catch the bee around its thorax and bristling forewings and send it buzzing back toward the sun? As I described in chapter 13, "Kermitronics," thought is not necessarily under our own control. Real enough experientially, upon investigation real agency may be nonexistent. Neither the mechanical causality of science nor the irreducible stochasticism of quantum flux, nor even the uncanny neo-Cartesian regulation of ourselves as second-order cybernetic machines,[7] is enough to save free will. We are still on that Uexküllian ride.

I argued in chapter 2, "Bataille's Sun and the Ethical Abyss," that life's problems come from solar excess: limited materials in a thermodynamically cycling system with evolving intelligence create participants who eventually recognize one another as gradients to be consumed. I don't see a way out of this situation. It seems, rather, constitutive of our existence, a violence, not so much of metaphysics, as Derrida might say, but of physics. The pedestrian fact is that on a planet of gradient-reducing beings building themselves from a limited material substrate, recycling becomes a necessity, and the most successful gradient reducers become among the most tempting treats. This does not obviate the possibility of compassion, however, the empathy that Karen Armstrong argues is the great common ground of world religions and which, as Saint Francis might add, includes nonhumans. Consider the *National Geographic* photographer Paul Nicklen who, visiting the Antarctic, dove into the ocean beneath the ice in an effort to photograph the elusive leopard seal, bigger than a grizzly bear. To his horror, Nicklen found what he was looking for. Immediately the leopard seal dropped the penguin that was in her mouth, turning toward him. The great mammal, engulfing Nicklen's headgear with her mouth—perhaps she mistook his head-mounted camera for the head of a strange new species—slowly lowered her giant teeth. Nicklen reports his legs shaking, his mouth dry, and other signs of intense fear. But then, to Nicklen's great relief, the seal, instead of bearing down and chopping off both head and camera, ended the threat display. She gently nuzzled him, like a dog play-biting; no skin was broken. But it was not over. Nicklen's new best friend went ashore onto the Antarctic wilderness and returned with a penguin for Nicklen. To the seal's surprise, Nicklen let the penguin shoot by him to safety. She tried again, to no avail. Assuming him to be unable to feed himself, she repeated her attempts at carnivorous gift-giving, bringing him smaller and weaker seals, partly consumed and dead seals. She appeared disgusted at what a useless predator he was, was desperate to feed him, and panicked that he might starve. This went on for four days. It was, Nicklen relates, an incredible experience. Such food sharing among members of the same and different species makes empathy and aid and spiritual belonging possible, even as it revolves around the sacrifice of excluded others. Excess provides not only for life's problems but also for its possibilities.

IN CHAPTER 8, "Thermosemiosis," I argued that behind our meaning making and goals is already always the in-itself-meaningless state of thermal equilibrium. In other essays I related this direction to psychedelic drugs, writing, and the recognition that human intelligence is already always a product of a far more encompassing, and in the end wiser, ecological intelligence. In chapter 1, "The Human Is More Than Human," developing the figure Stefan Helmreich calls *Homo microbians*,[8] I looked for symbiotic causes of our human moods, behaviors, and potential evolution. Although not explicit there, it is worth underlining that hypersex, the rampant gene-trading and endosymbiogenetics of microbes, depends on our status as open thermodynamic systems. Organisms, in the West long pictured as self-sufficient, Aristotelian or Linnaean species, as if they were creationistic instantiations of Platonic Ideas, are not. They are energetically and materially open systems. This is no minor point. It means that, when they get up close and microbial, in one another's faces and bodies, they have the potential to permanently merge. Scintillating solar light, with excess capacity for work, is degraded by trees, reaching toward the sun and aiming their leaves to maximize their capture of energy, most of which is used not for growth but for moving water through the stomata of leaves, which spreads the energetic equivalent per acre per summer of six tons of dynamite. People troll the world, laying cables and looking for petroleum and methane. Social insects feed and kill one another, regulating the temperature and humidity of their hives and hills and mounds. Cells merge, and cells made of cells merge, and organisms grown of merged organisms evolve into societies that verge on becoming organisms at a higher level of inclusiveness and organization once again. Life's modular basis, in other words, sets up the conditions for merging, for cellular economization, as organisms evolve ever-more-involuted solutions to ever-present problems of limited resources under regimes of excess and energy flow.

THE REINTRODUCTION OF FUNCTION—the reunion of ontos and telos with bios—is as empirically valid as it is spiritually salutary. As evolution connects us to other organisms, thermodynamics connects life to other complex systems. If life has been forged out of the clay of interstellar space, so thermodynamics instantiates a cosmic process of telic gradient reduction. The global system is organismic, and increases in energy efficiency, as well as in perception, add to the biosphere's ability to seek out and stably reduce cosmic and earthly gradients. This broad thermodynamics of life and its telic roots is, I maintain, after Copernican heliocentrism,

Darwinian evolution, and Wöhlerian antivitalism, a fourth Copernican-level deconstruction. Just as vitalism is wrong, and the chemical *stuff* of our bodies is not special, so mechanism is wrong and the basic *process* in which we are involved is not special. Thermodynamics is not all-powerful. Everything need not be reduced to or explained solely in terms of it. Telothermy is more like gravity, whose universal application in no way stops birds (whose wing energy comes from mitochondria-studded muscles powered by a redox, or delayed solar gradient) from flying.

But if we were cosmic journalists nursing our floating martinis and looking at old Io from the comfortable cushion of our space station lounge, and we wanted to report early-twenty-first-century humanity's basic understanding of itself, we could do worse than to answer the five basic questions of that propagandistically pliable medium: who, what, where, when, and why we are. We are organized collections of cosmically abundant atoms, living on a little planet in the inner part of a planetary system, engaged in a planetwide process of intrinsically purposeful energy distribution.

In other words, telic thermodynamics, as Charles P. Snow suggested about the second law, deserves wide understanding among all human beings interested in empirical science and their connection to the natural world. Paraphrasing Darwin, we might say that anyone whose disposition leads him or her to attach more weight to unexplained difficulties than to the explanation of facts will certainly not be attracted to the discovery of a vast overlap between human and living purpose and the telic tendencies of nonliving nature.

Although scientifically telic thermodynamics frames Gaia and symbiogenesis, it also illuminates other areas. For example, in *Beyond the Pleasure Principle* Sigmund Freud postulated a death drive. Life was an aberration in terms of inanimate matter; it thus unconsciously longed to return to its former state, to nonexistence.

The drives, Freud wrote, far from tending toward progress, seek "to reach an ancient goal by paths alike old and new. Moreover it is possible to specify this final goal of all organic striving. It would be in contradiction to the conservative nature of the instincts if the goal of life were the state of things which had never been attained. On the contrary, it must be an old state of things, an initial state from which the living entity has at one time or another departed and to which it is striving to return by the circuitous paths along which its development leads. If we are to take it as a truth that knows no exception that everything living dies for internal reasons—becomes inorganic once again—then we shall be compelled to say that 'the aim of all life is death' and, looking backwards, that 'inanimate things existed before living ones.'"

But Freud was perplexed not only at how the sex (and life) drive, eros or libido, could exist side by side with his death *Trieb* but also at what it was the sex drive was trying to recapture: "What is the important event in the development of living substance which is being repeated in sexual reproduction, or its forerunner, the conjugation of two protista? We cannot say; and we should consequently feel relieved if the whole structure of our argument turned out to be mistaken. The opposition between the . . . death instincts and the sexual or life instincts would then cease to hold."[9]

There is, I think, a simple, nonmetaphysical answer to Freud. Microbe studies, including organisms that have sex only when they are in dire straits and out of nitrogen, suggest fertilization began in multiple lineages among starved microbes, as discussed in chapter 5. If we grant (and it is not hugely generous) a primordial phenomenology to microbes that includes awareness and touch, urges and pains, chemical sensation and some sort of gratification, we can see the roots of a branching ur-drive. The phenomenological correlate to telos is a very basic kind of hunger, which accompanies or leads to destruction of local gradients needed to sustain the gradient-reducing system. Destructive hunger and procreative lust share a common root in gradient reduction.

In this context we might think of Freud's death drive, a danger to what Derrida wants to say is an archival drive of conservation (which might be allied with the Schopenhauerian will to live, although in a more symbolic register), in terms of an "archive fever . . . [a] limit [which is not] one limit or one suffering of memory among others: enlisting the in-finite, archive fever verges on radical evil."[10]

Of course, Derrida is interested in the necessary obfuscations, a kind of philosophical refusal to collapse the wave function, to keep open possibilities and opposites as part of his own deconstructive kabbalistic project, his metaphysical agenda, his fetish of the secret, the promise, and an ever-changing algebraic shibboleth, one might say the magic word, that protects as sacred an activity beyond discrete formalization. As part of a deep project of what might be called an apophatic or general Judaism, Derrida, critical of biologism, nonetheless gives Freud his materialistic due, underscoring that Freud's explicit distinction of archaeological and "impressive" representations of memory from the actual, spatial, material operations of the brain does not preclude that such possibilities (and by extension, other cognates of Freudian speculation) will occur in "the future of science.[11] It is as if Derrida wants to maintain (not negating the scientific register, but certainly not preferring it either) that life *in its bodies and in writing* is safeguarding a great secret (as if from itself) that is already always in as much danger of being destroyed as it is of being revealed.

Eros and thanatos stem from the common root of a *psychrotropos*—a movement toward cold (*psychros*), a dance of the atoms in chemically cycling and near-equilibrium systems, a natural movement to come, if not to order, then toward equilibrium. And with differences in effectiveness basically correlated to the maturity of subjectively judged beautiful ecosystems—rain forests and old-growth forests more effective than grasslands, more effective than deserts, and so forth—global life, 3.8 billion years strong with no allegiance to humanity, is the most effective dissipative planetary structure we know of. Borneo and Java jungles, for example, dissipate heat equivalent to Siberia in midwinter. This is known from thermal satellite measurements that show what superficially seems the reverse: these jungles are as cold as Siberia in midwinter. The area above them stays cool as latent heat dissipates via evaporation, water recondensing as cloud cover and rain in the atmosphere. The actively cycling rain forests, reminiscent of "purposeful" convection, are the most powerful biological dissipative systems on the planet. Highly species-rich, they also seem to have more long-term potential stability than the human technological monoculture that is tearing them down.

I have argued that both inanimate, near-equilibrium systems and technoscientific nation-states reflect the natural if amoral intelligence of a universe that creates complex systems to do its thermodynamic dirty work. Death, built into bodies as aging via apoptotic, insulin-regulated, telemorase-rationing cells, helps life hone its gradient-reducing function via natural selection, which works not only at the species but at the cellular (e.g., embryogenesis) and neurological (e.g., memes, mnemic algorithms, habits and their neural correlates) levels.

"I WAS A HIDDEN TREASURE and I longed to be known," says Allah in Sufi scripture.[12] Long before Renaissance paintings portrayed cherubic angels flying toward sun-spangled clouds, life was attracted to concentrated sources of energy, moving toward the light. *Tikkun olam* (תיקון עולם) is a Hebrew phrase meaning "repairing the world," originating in classical rabbinic literature but popularized by the sixteenth-century kabbalist Isaac Luria. As the universe continues to explode, we generally move away from radiating stars even as their matter and energy has, as us, come to life. In the material creation as described in the kabbalah, the world is infused with sparks of divine light, and the spirit's journey is to reunite with its divine source to restore the primordial realm. Here then is a potential story of the spirit in no obvious conflict with the scientific facts. Life and its intelligence, including human technological civilization, serve to equilibrate the environment, becoming enlightened as light itself dissipates.

We can sign on to the possibility, recognize the phenomenology of a destructive drive intrinsic to nature, destroying also information and archives, without dismissing the possibility that it has another, "deeper" meaning, although what this might mean in terms of a Gnosticism, say, is as quixotic as it would be for a Judeo-Derrideanism. It is worth mentioning, perhaps, that deconstruction is itself, both in its name (derived from Heideggerian *destruktion*) and in its "method," a form of destruction, of taking apart, of textual dismantling, rearranging, and suturing, a process reminiscent of microbial recycling to recapture, reassimilate, and reframe nutrients and chemical elements, a critical breakdown and return, the better to recapture prematurely foreclosed possibilities.

Derrida tells us (in *Archive Fever*) that *arkhe* is a kind of opposite of telos, meaning both commencement and commandment. The arche is the beginning, both literally and factually, and as a kind of fiat by the record keepers, this root word of archive and archon and archaeological coming from the Greek *arkheion,* "initially a house, a domicile, an address, the residence of the superior magistrates."[13] After I read my essay

"The Human Is More Than Human" to the American Anthropological Association in Montreal on the eve of my mother's death, Kim TallBear, a Native American theorist and activist, remained frankly critical of the very idea of origin.[14] Thus the *arkhe* not only as innocent commencement but as legislative fiat, a commandment that we recognize in a sense as arbitrary, fateful, self-foundational, with an element of fantasy to it commensurate with the judicial humorlessness of its authority. If the future is not closed, why should the past be? For Derrida and his ilk, the past is not original but originary, and the stakes of its foundationalism are too high to rush to agreement on beginnings so quixotic. So although I have said much about where we are going, and how, it would, in the end, let alone this conclusion, be presumptuous to pretend really to know.

I'VE HAD CAUSE before in this book to mention Niels Bohr's provocative quip, that the opposite of a trivial truth is wrong, but the opposite of a great truth is also a great truth. Perhaps an answer to the question of the purpose of life is subject to this dual logic. On the one hand, it seems impossible that our existence here doesn't have some purpose, however obscure. Beneath the veneer of the ordinary could there not *still* lurk something transcendental, metaphysical, sublime, a nacreous lining to the cosmic farce?

Isn't that the case? Or could this all be—as the out-and-out nihilists suggest—for nothing?

Was the whole idea of finding a purpose to life—other than the cybernetic existential solution, that is, the teleological tautology that life's purpose is to find its own purpose (featured in that the best-selling DVD *The Secret,* a kind of new age meditation on the power of realistic prayer)—or alternatively, the biological tautology, that life's purpose is merely to survive, or again, the verbal tautology that the purpose of life is a life of purpose, all just a bunch of BS?

In his unpublished paper, Carranza quotes Bruce Scofield, who continued, after my mother's death, her signature course, Environmental Evolution. Scofield, author of several hiking guides and an expert in Mayan astronomy who researches the possible scientific bases for some aspects of astrology, has pointed out that in the last 2,500 years of scientific history the current Newtonian–Darwinian mechanical model has been a relative rarity compared with organicist notions.[15]

What happens if we take seriously the notion that Earth is alive, or that it is conscious through us, or that the cosmos is?

One thing we can say is that only a universe that is illusorily separated from itself can be aware that it exists.

The perceiver and perceived must be separated for perception to take place. The breaking of the spell of this difference may be the unmediated unspeakable truth at the center of world mysticism, of perennial philosophy. Perception requires separation, spatiotemporality, the illusion of a subject–object difference. I call it "the paradox of aesthetic ontology."

To see anything we must think we are, in some way, different from what we perceive. We must hallucinate we are a part.

IN *AN APPRENTICESHIP OR THE BOOK OF DELIGHTS*, the brilliant and bizarre, lovely Brazilian writer Clarice Lispector describes her protagonist's Spinozistic awakening: "Lori had gone from the religion of her childhood to a nonreligion and now on to something less limited. She had reached the point where she believed in a God so vast that He was the Universe with all its galaxies. She had realized this the day before when she walked into the deserted ocean all alone. And because of His impersonal vastness He was not a God to Whom one could call out. What one could do was become part of Him and be great too."

ACKNOWLEDGMENTS

MANY THANKS to Tori Alexander, Nora Bateson, Wendell Berry, Dianne Bilyak, Dan Born, Eric Brado, Martin Brasier, Joanna Bybee, Joseph Cami, Carlos de Castro Carranza, Michael J. Chapman, Bruno Clarke, Paul Cobley, Trey Conner, Cristoph Cox, Kathryn Denning, Jacques Derrida, Rich Doyle, Celia Farber, Don Favareau, Stephan Harding, Peter Harries-Jones, John Hartigan, Stefan Helmreich, J. Marvin Herndon, Myra Hird, Bill Huth, J. Richard Kamber, Jean-Marie Kauth, Kermit, Andre Khalil, Eben Kirksey, Kalevi Kull, Frank L. Lambert, Timothy Leary, James Lovelock, James MacAllister, Ximena de la Macorra, Jaymie Matthews, Margaret McFall-Ngai, Josh Mitteldorf, Meghan Murphy, Michelle Murphy, Natasha Myers, Sina Najafi, Boris Petrov, Rick Ryals, Tonio Sagan, Stanley Salthe, Jan Sapp, Eric D. Schneider, Astrid Schraeder, John Scythes, Steve Shavel, Isabelle Stengers, Joy Stocke, F. J. R. "Max" Taylor, Rebecca Todd, Evan Thompson, William Irwin Thompson, Giuseppe Trautteur, Jessica Whiteside, Don Whitfield, Cary Wolfe. To all of these, none of whom did less than offer me a kind word (Wendell Berry), call me one of his heroes (Timothy Leary), or return a letter (Jacques Derrida), I am indebted. Others taught, inspired, shared with, edited, encouraged, supported, and lectured me.

Joy Stocke, Sina Najafi, Victoria N. Alexander, Frank L. Lambert, J. Marvin Herndon, Natasha Myers, and Rich Doyle provided detailed feedback on specific sections. So did Jason Weidemann of the University of Minnesota Press, who provocatively remarked that book writing is a mysterious mix of "crochet and alchemy," bon mots I heard as "*croquet* and alchemy," a near homonym but distinct algorithm that I followed for most of the book but whose perhaps wildly different consequences I have not had time to redress. Jason gave me the Apache art critic Paul Chaat Smith's slim book *Everything You Know about Indians Is Wrong* as a model. I loved the line in Smith's acknowledgments where he agrees to share not just the credit but the blame with some of his friends for any mistakes he may have made. I concur wholeheartedly and extend a similar acknowledgment across the literary nethersphere in appreciation for this plagiarizable observation. Thank you all.

NOTES

INTRODUCTION

1. Sam Harris, *Free Will* (New York: Free Press, 2012), 64.

2. Rick Warren, *The Purpose-Driven Life: What on Earth Am I Here For?* (Grand Rapids, Mich.: Zondervan, 2002), 20.

3. Martin Heidegger, *The Essence of Human Freedom: An Introduction to Philosophy* (London: Continuum, 2002), 5.

4. Ibid.

5. Ibid., 205.

6. P. H. Barrett, "Early Writings of Charles Darwin," in *Darwin on Man: A Psychological Study of Scientific Creativity; together with Darwin's Early and Unpublished Notebooks,* edited by H. E. Gruber (London: Wildwood House, 1974).

7. C. R. Darwin, *On the Origin of Species by Means of Natural Selection, or the Preservation of Favoured Races in the Struggle for Life* (London: John Murray, 1859), 482.

8. Sharon E. Kingsland, "The Beauty of the World: Evelyn Hutchinson's Vision of Science," in *The Art of Ecology: Writings of G. Evelyn Hutchinson*, edited by David K. Skelly, David M. Post, and Melinda D. Smith (New Haven, Conn.: Yale University Press, 2010), 4.

9. Ibid.

1. THE HUMAN IS MORE THAN HUMAN

1. Richard Feynman, *The Meaning of It All: Thoughts of a Citizen-Scientist* (New York: Basic Books, 1999), 10.

2. Karen Barad, *Meeting the Universe Halfway: Quantum Physics and the Entanglement of Matter and Meaning* (Durham, N.C.: Duke University Press, 2007).

3. Freeman Dyson, *Origins of Life* (Cambridge: Cambridge University Press, 1999).

4. Clair Folsome, "Microbes," in *The Biosphere Catalogue,* edited by Tango P. Snyder (Fort Worth, Tex.: Synergetic, 1985), 51–56.

5. Steven L. Salzberg, Owen White, Jeremy Peterson, and Jonathan A. Eisen, "Microbial Genes in the Human Genome: Lateral Transfer or Gene Loss?" *Science* 292 (2001): 1903–6.

6. Junjie Qin et al., "A Human Gut Microbial Gene Catalogue Established by Metagenomic Sequencing," *Nature,* March 4, 2010, 59–65.

7. Yun Kyung Lee and Sarkis K. Mazmanian, "Has the Microbiota Played a Critical Role in the Evolution of the Adaptive Immune System?" *Science,* December 24, 2010, 1768–73; and see Gérard Eberl, "A New Vision of Immunity: Homeostasis of the Superorganism," *Mucosal Immunology,* May 5, 2010, 450–60.

8. In *What Is Sex?* "I" put forward the notion of *hypersex*—the great realm of gene trading not down through the generations but from organism to organism, across types and species, sometimes as we've recently seen in the news, as decisions of what to eat, as live food also contains genes. The distributed desire that accompanies organisms commingling, that leads them to eat, meet, engulf, invade, trade genes, acquire genomes, and sometimes permanently merge is bigger than eros; it is more like some primordial cosmic wanderlust. In Myra Hird's book *The Origins of Sociable Life* I see hypersex refracted back at me, as it were, in fascinating new ways that include not only an evocation of the power of symbiogenesis—the evolution of new species by symbiosis—but in a cross-fertilization of interdisciplinary thought and fields. Imagining that some readers may presume she suffers from microbiology or geosciences envy, Hird, who observed the laboratory of my mother, Lynn Margulis, and whose book I read in preparation for this chapter, admits it but writes: "Might this castigation . . . be a ruse to dismiss further critical reflection? I worry that a sense of smugness pervades the social sciences generally and licenses the false impression that natural scientists are largely ignorant of philosophical and social studies of science (they/scientists are observed; we/social scientists are observers) while we can proceed with social scientific analyses that assume we may gain sufficient understanding of phenomena by studying what we distinguish as social aspects of materiality" (Myra J. Hird, *The Origins of Sociable Life: Evolution after Science Studies* [Basingstoke: Palgrave Macmillan, 2009], 181). The very assumption that we can be completely on the outside of what we are observing (which is subtly and ironically reinforced by making this assumption explicit) is a recurrent problem, salient in the study of science, what might be called "the meta-objective aporia": that because we are looking at people's frames, their hidden assumptions and sociocultural context, we are not equally mired in our own.

9. Margaret McFall-Ngai, "Origins of the Immune System," in *Chimeras and Consciousness: Evolution of the Sensory Self,* edited by Lynn Margulis, Celeste A. Asikainen, and Wolfgang E. Krumbein (Cambridge, Mass.: MIT Press, 2011).

10. Dick Teresi, "Lynn Margulis Says She's Not Controversial, She's Right: It's the Neo-Darwinists, Population Geneticists, AIDS Researchers, and English-Speaking Biologists as a Whole Who Have It All Wrong," *Discover,* April 2011, 66–71.

11. Christopher Carter, "The Human Genome Is Composed of Viral DNA: Viral Homologues of the Protein Products Cause Alzheimer's Disease and Others via Autoimmune Mechanisms," Nature Precedings, 2010, http://precedings.nature.com/documents/4765/version/1.

12. Carl Zimmer, "Tending the Body's Microbial Garden," *New York Times,* June 19, 2012.

13. Carrie Arnold, "Gut Microbes May Drive Evolution: The Bacteria That Live Quietly in Our Bodies May Have a Hand in Shaping Evolution," *Scientific American* (2012), http://www.scientificamerican.com/article.cfm?id=backseat-drivers (published in print as "Backseat Drivers") (accessed April 29, 2012).

14. Karen Milius, "Green Sea Slug Is Part Animal, Part Plant," *Science News* (2011), Wired Science, http://www.wired.com/wiredscience/2010/01/green-sea-slug/.

15. Valerie Brown, "Bacteria 'R' Us," *Miller-McCune* (2010), http://www.miller-mccune.com/science-environment/bacteria-r-us-23628/.

16. Jaroslav Flegr, Jiří Klose, Martina Novotná, Miroslava Berenreitterová, and Jan Havlíček, "Increased Incidence of Traffic Accidents in Toxoplasma-Infected Military Drivers and Protective Effect RhD Molecule Revealed by a Large-Scale Prospective Cohort Study," *BMC Infectious Diseases* 9 (2009): 72.

17. Kathleen McAuliffe, "How Your Cat Is Making You Crazy," *Atlantic Magazine,* March 2012, http://www.theatlantic.com/magazine/archive/2012/03/how-your-cat-is-making-you-crazy/8873/.

18. Dorion Sagan and Lynn Margulis, "Candidiasis and the Origin of Clowns," *New England Watershed,* October 2005, 16–19. Reprinted in Margulis and Sagan, *Dazzle Gradually: Reflections on the Nature of Nature* (White River Junction, Vt.: Chelsea Green), 146–52.

19. This chapter originally was given as an address in 2011 in Montreal to the Society of Cultural Anthropology.

20. Alfred North Whitehead, *Science and the Modern World* (New York: New American Library, 1962), 10.

21. Ibid., 15.

22. Ibid., 17.

23. These are the rough facts of thermodynamics that I suggest sum to a fourth Copernican deconstruction. (The first is heliocentrism. The second is organicism: we are not made of any

rare, special stuff, as the vitalists thought, but from organic compounds, cosmically abundant and found in space. The third is Darwin's emplacement of us within a temporal continuum, not above but within the animals.) The basic facts of the fourth deconstruction, which shows that the process of life, no less than its material constituents, is common, are these:

1. We live in an energetic cosmos where energy, if not hindered, tends to spread. Though simple, this statement is in fact a modern statement (which applies to open as well as sealed systems) of the second law of thermodynamics. Entropy is not mysterious but simply a mathematical abstraction (a ratio) that measures the spread of energy.

2. Although Ludwig Boltzmann initially identified entropy with disorder, the natural tendency for energy to spread is often accomplished by organized, self-like systems. Energy's spread, or the reduction of gradients, is accomplished more effectively by whirlpools and storms, recursive chemical reactions, and other cyclical processes than by unorganized matter. More entropy is produced, more gradients are reduced, and more energy is dissipated by such natural "self-like" organizations.

3. Thus not only does life not *violate* the second law, but living matter actively accomplishes a basic physical task implicit in the basic structure of matter in disequilibrium, which is inherently teleological, tending toward an end (at least at the local scale) of equilibrium and employing complex systems to accomplish those ends. Metabolizing organisms spread more energy than nonarranged matter, and the spread of growing, reproducing life better accomplishes the second law–described mandate of energy to spread (gradients to be reduced), which is at the core of thermal physics. Einstein identified thermodynamics as, of all the areas of physics, the one least likely to change in the future. In manifesting the second law, life is completely natural. Nonetheless, both scientists and religionists continue to miss this, perhaps because entropy has been conflated with disorder rather than recognized as a measure of energy's spread. For example, Pope Pius XII invoked the second law as proof of God's existence, apparently because only he could violate the law of ever-increasing disorder to produce organized life. But the neo-Darwinist philosopher Daniel Dennett exemplarily repeats this mistake when he writes that life forms "are things that defy" and constitute a "systematic reversal" of the second law (*Darwin's Dangerous Idea* [New York: Simon and Schuster, 1995], 69).

4. Despite the tendency for energy to spread manifested by life, this tendency is not *maximized* by life. To maximize your entropy production right now you would have to burst into flames. In fact, measurements show that fast-growing organisms such

as juvenile forms and pioneer species in early-stage ecosystems have higher specific entropy production or entropy production per volume than mature forms such as adults or climax-stage ecosystems. The nature of the universe, codified by the second law, tugs organisms to find ways to work together to stably reduce gradients, dissipating the energy that sustains them. As populations must always apportion a fragment of the energy they commandeer toward their own preservation, both the entropy maximization theorists within thermodynamics and the cultural theorists who critique them miss the brute empiricism of sustainability in real ecosystem dynamics.

5. Evolution trends. A universe marked by energy spread is implicitly telic, goal-directed, or end-driven ("preparing the way" for both life and consciousness). Even simple nonequilibrium systems unconsciously calculate ways to reach equilibrium, and complex systems maintain their nonequilibrium as they lay larger energy gradients to waste. The tendency of energy to spread confers a direction upon evolution as a whole. Here evolutionists as distinct politically as Richard Dawkins and Stephen Jay Gould both err (unintentionally providing fuel for creationists by ignoring clear evidence) when they characterize evolution as intrinsically random. In fact, evolution is accompanied by clear, measurable vectors: Evolution's main trends include increase in number of individuals, species, and taxa; increase in bacterial and animal respiration efficiency; increase in number of cell types; and long-term increases, despite periodic setbacks from mass extinctions, in global biodiversity, connected sentience, and aggregate information-processing abilities. Evolution, in other words, is a thermodynamic phenomenon. And it is not just theoretical. Using living representatives of animal groups and plotting a curve according to the order in which their groups appeared in the fossil record, the Russian scientist Alexander Zotin quantified a striking trend toward increase in oxygen efficiency over geological time ("Bioenergetic Trends of Evolutionary Progress of Organisms," in *Thermodynamics and Regulation of Biological Processes,* edited by Alexander I. Lamprecht [Berlin: de Gruyter, 1984]).

Another Russian scientist, Vladimir I. Vernadsky (whose work Bataille read in Paris in 1929), pointed out that the number of chemical elements in the Periodic Table that have become incorporated into life at Earth's surface has increased dramatically over evolutionary time. Later evolved respiring bacteria produce seventeen times more ATP from sugar molecules than fermenting bacteria. Using a gauge he calls "free energy density rate"—consisting of ergs, a measure of work (per second per gram)—the astrophysicist Eric Chaisson scores the human brain higher than any part of a known star. Using a slightly different measure (watts per kilogram), which engineers call "specific power," we find that the human body, thanks to the energy centers known as mitochondria, is also, adjusted for mass, moving energy through themselves at ten thousand times the rate of the sun. According to this gauge, basically a proxy for energy flow, Chaisson points to increases in complex systems as they evolve from galaxies through stars to biospheres, reptiles, mammals, brains, societies, and computers. Although it is not perfect (it is anomalously high in flames, hummingbirds, and respiring bacteria), Chaisson's free energy density rate matches our general intuition of complexity and underscores the crucial point that evolving complexity depends on energy flow through systems (*Cosmic Evolution: The Rise of Complexity in Nature* [Cambridge, Mass.: Harvard University Press, 2001]). Life would thus seem to be purposive, going somewhere, not "on its own" but along with the universe. The evidence countermands evolutionary prejudices in favor of randomness *and* religious predilections for a humanlike deity: Life is going somewhere, although its telos would seem, on the basis of the evidence, to be operating *through* rather than *to* man. I return to this question in a more vernacular style in the final essay.

24. Josh Mitteldorf and John Pepper, "Senescence as an Adaptation to Limit the Spread of Disease," *Journal of Theoretical Biology* 260, no. 2 (2009): 186–95.

25. Robert S. Loomis and David J. Connor, *Crop Ecology: Productivity and Management in Agricultural Systems* (Cambridge: Cambridge University Press, 1992).

26. Stefan Helmreich, *Alien Ocean: Anthropological Voyages in Microbial Seas* (Berkeley: University of California Press, 2009), 283.

2. BATAILLE'S SUN AND THE ETHICAL ABYSS

1. Dorion Sagan and Lynn Margulis, "Facing Nature," in *Biology, Ethics, and the Origins of Life*, edited by Holmes Rolston III (Boston: Jones and Bartlett, 1995), 39–62; Sagan and Margulis, "Gaia and the Ethical Abyss," in *The Good in Nature and Humanity: Connecting Science, Religion, and Spirituality with the Natural World,* edited by Stephen R. Kellert and Timothy J. Franham (New Haven, Conn.: Island Press, 2001), 91–101.

2. Monoculturalism is the slippery notion that there is one, scientifically agreeable upon world, discovered by a science that, although it developed in Europe, makes planetary citizens of us all. A seductive example is given by Carl Sagan: "How far will our nomadic species have wandered by the next century, and the next millennium? Our remote descendants, safely arrayed on many worlds through the solar system and beyond, will be unified, by their common heritage, by their regard for their home planet, and by the knowledge that whatever other life may be, the only humans in all the universe come from Earth. They will gaze up and strain to find the blue dot in their skies. They will marvel at how vulnerable the repository of all our potential once was, how perilous our infancy, how humble our beginnings, how many rivers

we had to cross before we found our way" (*Pale Blue Dot: A Vision of the Human Future in Space* [New York: Ballantine Books, 1997], 334). See Bruno Latour, "Whose Cosmos, Which Cosmopolitics? Comments on the Peace Terms of Ulrich Beck," *Common Knowledge* 10, no. 3 (2004): 454.

3. Alan Watts, *The Book: On the Taboo against Knowing Who You Are* (New York: Vintage, 1989), 4.

4. Josh Mitteldorf, "Aging Is Not a Process of Wear-and-Tear," *Rejuvenation Research* 13 (2010): 322–25; and see Mitteldorf, "Evolutionary Origins of Aging," in *Approaches to the Control of Aging: Building a Pathway to Human Life Extension*, edited by Gregory M. Fahy, Michael D. West, L. Stephen Coles, and Steven B. Harris (New York: Springer, 2010).

5. Ed Cohen, *A Body Worth Defending: Immunity, Biopolitics, and the Apotheosis of the Modern Body* (Durham, N.C.: Duke University Press, 2009), 1–2.

6. As Timothy Campbell says in the translator's introduction, "In any case, for the present discussion what matters most is that Derrida believes that September 11 cannot be thought independently of the figure of immunity; indeed, that as long as the United States continues to play the role of 'guarantor or guardian of the entire world order,' autoimmunitary aggression will continue, provoked in turn by future traumatizing events that may be far worse than September 11" (Roberto Esposito, *Bíos: Biopolitics and Philosophy*, translated by Timothy Campbell, Posthumanities 4 [Minneapolis: University of Minnesota Press, 2008], xviii). Esposito is interesting in part because he takes seriously the potential homology between the body and the body politic even while reviling its totalitarian logic: "From this perspective, we can say that immunization is a negative 'form' of the protection of life. It save, insures, and preserves the organism, either individual or collective, to which it pertains, but it does not do so directly, immediately, or frontally; on the contrary, it subjects the organism to a condition that simultaneously negates or reduces its power to expand. Just as in the medical practice of vaccinating the individual body, so the immunization of the political body functions similarly, introducing within it a fragment of the same pathogen from which it wants to protect itself" (*Bíos*, 46). Ironically, a liberal attitude to the Estonian ethologist, somewhat mirroring Adolf Hitler's vegetarianism, because the former granted such theoretical respect to the inner worlds of animals that he has been embraced by animal studies if not animal rights groups, Uexküll makes clear the terrifying logic of the biostate in his "*Staatsbiologie*, which was also published in 1920 by Baron Jakob von Uexküll with the symptomatic subtitle *Anatomy, Physiology, and Pathology of the State*. Here, as with Kjellen, the discourse revolves around the biological configuration of the state-body that is unified by harmonic relations of its own organs, representative of different professions and competencies. . . . Threatening the public health of the German body is a series of diseases, which obviously, referring to the revolutionary traumas

of the time, are located in subversive trade unionism, electoral democracy, and the right to strike: tumors that grow in the tissues of the state, causing anarchy and finally the state's dissolution. It would be 'as if the majority of the cells in our body (rather than those in our brain) decided which impulses to communicate to the nerves'" (*Bíos,* 17–18).

7. The engagement of the biopolitical is "medieval-microcosmic" in its unapologetic appropriation/recognition of the biological for political discourse. The equivalence brain:body; govt:body politic, like the equivalence immune system:body; national security forces:body politic is homologous not only in terms of cultural derivation and a pre-French revolutionary thinking to do with the divine right of kings, but also perhaps in terms of the natural development that is far easier for a social Darwinism/futurism/sociobiology/scientism/Nazism to embrace than traditional humanism and the "once burned, twice shy" mind-set of humanities and posthumanities from cultural anthropology/sociology/social studies of science to geography and so on. It is anathema to think the bios as always already more than human when it is always already mediated by human consciousness. This is a constitutive aporia that cannot be solved by a new agential cut. The impossible shape of {{{nature}culture}nature} (to put it in hierarchic notation with the outer brackets being the most inclusive) allies it to Paul Valéry's line that he lives in a world inside his head inside the world. Nonethe*less,* if we accept for the sake of argument that the three imploded buildings of September 11 were demolished as a provocation for "an endless cycle of violence" (as Dick Cheney put it), we can, with no need to identify the perpetrators in whole or part, interpret the phenomenon in terms of the autoimmunity of globalization. Compared with the demolishing of Hiroshima and Nagasaki that ended World War II, the destruction was nowhere so severe. But the importance for the global body politic was equally if not more profound: whoever did it, the eradication of the Twin Trade Towers initiated a global immune response. If it was a false flag operation, the latest and most spectacular of the nationalist master weapon of killing one's own to trigger systemic response, then we have a perfect conflation of the political and biological acts of autoimmunization. The catastrophe would be formally equivalent to a vaccination against the perceived dangers of spreading, suicide-friendly, Islamic terrorism.

It only requires an abdication of absolute morality to imagine the mind-set of a globalist supranational hegemony that would not be above faking Islamic, white power, and other attacks to preclude them, as well as to consolidate power using the Machiavellian "short cut" of engineered fear and crisis conditions. Fears of backlash against elite leaders, not to mention scenarios of mob mobilizations potentially worse than engineered terror, work to concretize the propagandistically useful notions of democracy, antisexism, antiracism, multiculturalism, and so on and embolden extralegal power checked only by its own equally deceptive, functionally criminal, and powerful cousins. False flags as a politico-military attractor can thus potentially be understood within the framework of realpolitik and "immunization"—you purposely stage a feared threat, taking the wind out of the sails of those who would do it to you. You make big bucks in the process, all the while congratulating yourself on your mastery of population control and applied Platonism. Shadia Drury argues that neocons learned from Leo Strauss, who argued that Plato's real mouthpiece is not Socrates but Thrasymachus, whom Strauss defends along with Machiavelli as political realists necessary to protect against the stupidity of the masses and reprisals against their secretive masters (Danny Postel, "Noble Lies and Perpetual War: Leo Strauss, the Neo-cons, and Iraq," 2003, http://www.informationclearinghouse.info/article5010.htm [accessed May 20, 2012]).

8. Whitehead, *Science and the Modern World*, 10.

9. Cohen, *A Body Worth Defending*, 281.

10. McFall-Ngai, "Origins of the Immune System."

11. Esposito, *Bíos*, 17.

12. Herbert Spencer, preface to *The Data of Ethics* (London: Williams and Norgate, 1879).

3. THE POST-MAN ALREADY ALWAYS RINGS TWICE

1. Otto Rank, *The Trauma of Birth* (New York: Harcourt, Brace, 1929).

2. Paul West, *Master Class: Scenes from a Fiction Workshop* (New York: Harcourt Brace, 2001).

3. Samuel Butler, *The Shrewsbury Edition of the Works of Samuel Butler*, edited by Henry Festing Jones and A. T. Bartholomew (New York: E. P. Dutton, 1923–26).

4. Alan Watts, *Does It Matter? Essays on Man's Relation to Materiality* (New York: Pantheon Books, 1970).

8. THERMOSEMIOSIS

1. Thomas Hager, "World's Greatest Invention," http://thomashager.net/haber-bosch-the-worlds-greatest-invention/ (accessed May 28, 2012).

2. A. S. Eddington, *The Nature of the Physical World* (Cambridge: Cambridge University Press, 1933), 74–75.

3. C. P. Snow, *The Two Cultures* (Cambridge: Cambridge University Press, 1998), 14–15.

4. Frank L. Lambert (b. 1918), a professor emeritus of Occidental College, Los Angeles, has since 1999 produced publications that have led to twenty-nine chemistry textbooks, reaching approximately 450,000 students (to date), to discard "disorder" as a description of entropy. See, for example, Frank L. Lambert, "Disorder—a Cracked Crutch for Supporting

Entropy Discussions," *Journal of Chemical Education* 79 (2002): 187–92; Lambert, "Shuffled Cards, Messy Desks, and Disorderly Dorm Rooms—Examples of Entropy Increase? Nonsense!" *Journal of Chemical Education* 76 (1999): 1385–87; and Lambert, "Entropy Is Simple, Qualitatively," *Journal of Chemical Education* 79 (2002): 1241–46. His websites include secondlaw, shakespeare2ndlaw (where he discusses C. P. Snow), and entropysite, all with the suffix *oxy.edu*; the latter containing multiple links to relevant publications at a variety of levels.

5. Dennett, *Darwin's Dangerous Idea,* 69.

6. Where ΔS is change in entropy, dq is change in energy, and T is temperature.

7. Ludwig Boltzmann, *Lectures on Gas Theory* (1898), 443.

8. Myron Tribus and Edward C. McIrvine, "Energy and Information," *Scientific American,* September 1971, 224.

9. Frank L. Lambert, "The Conceptual Meaning of Thermodynamic Entropy in the Twenty-First Century," *International Research Journal of Pure and Applied Chemistry* 1, no. 3 (2011): 67, http://www.sciencedomain.org/issue.php?iid=82&id=7.

10. Steve Adams, "No Way Back," *New Scientist,* October 22, 1994, 1–4.

11. Frank Lambert, "Disorder in Thermodynamic Entropy," http://entropysite.oxy.edu/boltzmann.html (accessed June 22, 2012).

12. Harvey Leff, "Thermodynamics Is Easy—I've Learned It Many Times," *Physics Teacher* 45 (February 2007): 71–72. That initial brief article has now been very thoroughly supported by a series of five articles for physics instructors that are accessible online, http://www.csupomona.edu/~hsleff/selpubs.html.

13. Isabelle Stengers, *Cosmopolitics I, In the Name of the Arrow of Time: Prigogine's Challenge* (Minneapolis: University of Minnesota Press, 2010), 24.

14. P. M. Harman, ed., *The Scientific Letters and Papers of James Clerk Maxwell,* vol. 2 (Cambridge: Cambridge University Press, 1995), 947.

15. Jesper Hoffmeyer, *Signs of Meaning in the Universe* (Bloomington: University of Indiana Press, 1997).

16. Buckminster Fuller, *I Seem to Be a Verb: Environment and Man's Future* (New York: Bantam Books, 1970).

17. Giorgio Agamben, *The Open: Man and Animal,* translated by Kevin Attell (Palo Alto, Calif.: Stanford University Press, 2003).

18. Wendy Wheeler, *The Whole Creature: Complexity, Biosemiotics, and the Evolution of Culture* (London: Lawrence and Wishart, 2006).

19. Agamben, *The Open.*

20. Candace B. Pert, *Molecules of Emotion: The Science behind Mind–Body Medicine* (New York: Simon and Schuster, 1999).

21. Josh Mitteldorf and John Pepper, "How Can Evolutionary Theory Accommodate Recent Empirical Results on Organismal Senescence?" *Theory in Biosciences* 126, no. 1 (2007): 3–8.

22. Jesper Hoffmeyer, "Semiotic Freedom," in *Information and the Nature of Reality: from Physics to Metaphysics,* edited by Paul Davis and Niels Henrik Gregersen (Cambridge: Cambridge University Press, 2010), 185–204.

23. Dorion Sagan, "Thermodynamics and Thought," in Margulis, Asikainen, and Krumbein, *Chimeras and Consciousness,* 241–50.

24. Dorion Sagan and Jessica Whiteside, "Thermodynamics and the Purpose of Life," in *Scientists Debate Gaia: The Next Century,* edited by Stephen H. Schneider, James R. Miller, Eileen Crist, and Penelope J. Boston (Cambridge, Mass.: MIT Press, 2004), 173–86.

25. Eliseo Fernández, "Energy, Semiosis, and Emergence—the Place of Biosemiotics in an Evolutionary Conception of Nature," Eleventh Annual International Gathering in Biosemiotics, Dactyl Foundation, New York City, June 21–26, 2011.

26. Peter Brooks, *Reading for the Plot: Design and Intention in Narrative* (Cambridge, Mass.: Harvard University Press, 1992).

9. LIFE GAVE EARTH THE BLUES

1. A. V. Lapo, *Traces of Bygone Biospheres,* translated by V. Purto (Moscow: Mir, 1982). See also "Mining Nickel the Easy Way," *New Scientist,* March 16, 1972, 583.

10. MOUSETRAP

1. Arthur Koestler, *The Ghost in the Machine* (Washington, D.C.: Henry Regnery, 1971), 158.

2. Robert Rosen, *Life Itself: A Comprehensive Inquiry into the Nature, Origin, and Fabrication of Life* (New York: Columbia University Press, 1991).

3. Carl Sagan, *The Cosmic Connection: An Extraterrestrial Perspective* (New York: Dell, 1975), 150.

4. "Teleology is like a mistress to a biologist," quipped the British geneticist J. B. S. Haldane. "He cannot live without her but he's unwilling to be seen with her in public." Aware of the problem, the evolutionist Ernst Mayr divided purpose into three kinds: "teleomatic," which refers to "end-producing" forces like gravity, which brings a marble rolling in a bowl until it wobbles to a stop in a concave depression; "teleonomy," which refers to "end-directed" processes produced by natural selection, like the heart, which acts to pump blood; and "teleology" proper, which for Mayr is conscious and willful. Mayr took the middle term *teleonomic,* from

Colin Pittendrigh, who wrote back to him in 1970: "You ask about the word 'teleonomy.' You are correct that I did introduce the term into biology and, moreover, I invented it. . . . Organization is more than mere order; order lacks end-directedness; organization is end-directed. . . . The most general of all biological 'ends, or 'purposes' is of course perpetuation by reproduction." This is incorrect, however, as thermodynamic equilibrium, which metabolism, growth, and reproduction perpetuate, is more general. In 1982 the historian of science David Hull wrote, "Today the mistress has become a lawfully wedded wife. Biologists no longer feel obligated to apologize for their use of teleological language. . . . The only concession which they make to its disreputable past is to rename it 'teleonomy.'" In retrospect the taboo against purpose was an allergic reaction against the faintest whiff of Aristotelian teleology, which was conflated by the church with a humanlike God. But it is unwarranted. Natural complex systems are cyclical, so they can be their own purposes. But beyond maintenance and reproduction is the tendency toward equilibrium. The organization and selves of self-organizing systems do not exist without the "push" coming from the telic tendencies described in the second law. And that's no BS.

5. Sze Zeng, "Who Is Stanley N. Salthe; What Did He Do and Say?" http://szezeng.blogspot.com/2010/11/who-is-stanley-n-salthe-what-did-he-do.html (accessed October 28, 2012).

11. PRIESTS OF THE MODERN AGE

1. For example, the last two sentences of *On the Origin of Species* in the second through sixth and last editions: "Thus, from the war of nature, from famine and death, the most exalted object which we are capable of conceiving, namely, the production of the higher animals, directly follows. There is grandeur in this view of life, with its several powers, having been originally breathed by the Creator into a few forms or into one; and that, whilst this planet has gone cycling on according to the fixed law of gravity, from so simple a beginning endless forms most beautiful and most wonderful have been, and are being evolved."

2. Galileo Galilei, *Il Saggiatore* (Rome, 1623); Galilei, *The Assayer*, translated by Drake Stillman and C. D. O'Malley, in *The Controversy on the Comets of 1618* (Philadelphia: University of Pennsylvania Press, 1960).

3. Galileo Galilei, *Dialogue concerning the Two Chief World Systems* (1632), 87–88; Italo Calvino, *Why Read the Classics?* (New York: Pantheon, 1999), 199.

4. David Bohm, *On Creativity* (New York: Routledge, 2004).

5. James A. Shapiro, "What Is the Key to a Realistic Theory of Evolution?" Huffington Post, www.huffingtonpost.com/james-a-shapiro/what-is-the-key-to-a-real_b_1280685.html?ref=science (accessed February 19, 2012).

6. Jerry Coyne, "A Colleague Wrongfully Disses Modern Evolutionary Theory," Why Evolution Is True, 2012, http://whyevolutionistrue.wordpress.com/2012/02/18/a-colleague-wrongfully-disses-modern-evolutionary-theory/ (accessed February 19, 2012).

7. Lingua Franca, *The Sokal Hoax: The Sham That Shook the Academy* (Lincoln, Neb.: Bison Books, 2000).

8. As attested to when, in the late 1890s, the minister of posts and telegraphs, to whom he applied for funding, did not answer him but wrote "to the Longara"—which was the name of the insane asylum in Rome—on his application.

9. Ludwik Fleck, *Genesis and Development of a Scientific Fact* (Chicago: University of Chicago Press, 1979). Fleck (1896–1961) wrote this book before the better-known meditations on the making of science facts by Thomas Kuhn, who acknowledged his debt to Fleck.

10. Lynn Margulis, "Origin of Evolutionary Novelty by Symbiogenesis," in *Biological Evolution: Facts and Theories: A Critical Appraisal of 150 Years after "The Origin of Species,"* edited by G. Auletta, M. Leclerc, and R. A. Martinez (Rome: Gregorian and Biblical Press, Piazza della Pilotta, 2011), 107–16. See also "The Human Is More Than Human," chapter 1 of this volume.

11. Jerry Coyne, "Lynn Margulis Disses Evolution in Discover Magazine, Embarrasses Both Herself and the Field," Why Evolution Is True, 2011, http://whyevolutionistrue.wordpress.com/2011/04/12/lynn-margulis-disses-evolution-in-discover-magazine-embarrasses-both-herself-and-the-field/ (accessed December 5, 2012).

12. "Coauthorial Critique" of *Acquiring Genomes: A Theory of the Origins of Species* (Cary, N.C.: Basic Books, 2002), http://www.amazon.com/review/R2VM3EHYAOYP2D.

13. Karl Popper, *Unended Quest: An Intellectual Autobiography* (La Salle, Ill.: Open Court Publishing, 1982), 33–34.

14. J. Marvin Herndon, *Indivisible Earth: Consequences of Earth's Early Formation as a Jupiter-Like Gas Giant* (Cary, N.C.: Thinker Media, 2012); Herndon, *Origin of the Geomagnetic Field: Consequence of Earth's Early Formation as a Jupiter-Like Gas Giant* (Cary, N.C.: Thinker Media, 2012); Herndon, *Beyond Plate Tectonics: Consequence of Earth's Early Formation as a Jupiter-Like Gas Giant* (Cary, N.C.: Thinker Media, 2012).

15. Herndon, *Indivisible Earth*; Herndon, *Origin of the Geomagnetic Field.*

16. The Kamland Collaboration, "Partial Radiogenic Heat Model for Earth Revealed by Geoneutrino Measurements," *Nature Geoscience*, July 17, 2011, http://www.nature.com/ngeo/journal/v4/n9/full/ngeo1205.html.

17. J. Marvin Herndon, "Corruption of Science in America," *Dot Connector Magazine* 2, no. 14 (2011): 23–30.

18. See, for example, http://www.youtube.com/watch?v=BwgmzbnckII&feature=related, which has been criticized as AIDS denialism; but remember questioning the HIV-AIDs connec-

tion is not the same thing as denying AIDS; see also the WikiLeaks cable from 2009 from the Office of the Secretary of the State of the United States that HIV infection is "not a communicable disease that is of significant public health risk": http://truthbarrier.com/2012/03/14/wikileaks-cable-us-gov-ceases-hiv-testing-visa-applicants-and-calls-hiv-infection-not-a-communicable-disease-that-is-of-significant-public-health-risk/. Contributing to P. Z. Myers's "Pharyngula" blog in 2007, Lynn Margulis wrote: "Peter Duesberg is a fine scientist, I have read his book and examined some of the scientific papers upon which it is based. From the CDC (Center for Disease Control) in Atlanta I have requested the scientific papers that prove the causal relationship between the HIV retrovirus and the IMMUNODEFICIENCY SYNDROME commonly known as AIDS. They have never sent even references to the peer-reviewed primary scientific literature that establishes the causal relationship because they can't. Such papers do not exist. I have seen all four of the films made by Coleman Jones and colleagues in Toronto. Film #3 in the series is most telling. Although no strong evidence exists for any simple causal relationship what is clear is that the HIV claim is erroneous by the standards of microbiology and virology. . . . I heard a talk by a 'medical scientist' from the Harvard Medical School at a meeting at Roger Williams Univ in Rhode Island . . . who attempts to design an HIV vaccine. He claimed the HIV virus mutates a billion times in 48 hours. It became clear that the HIV virus has no clear identity. . . . One can be more honest if the earliest stages of evolution are the objects of study. And this way I can lay low and not be 'name-called' (i.e., 'denialist') because I ask hard questions and require solid evidence before I embrace a particular causal hypothesis. Indeed, is not my attitude of inquiry exactly what science is about?" (http://scienceblogs.com/pharyngula/2007/03/lynn_margulis_blog_tour.php).

19. Guy Gugliotta, "Is Earth's Core a Nuclear Fission Reactor?" *Washington Post,* March 24, 2003, 6.

20. "Science Guardian: Paradigms and Power in Science and Society," http://www.scienceguardian.com/blog/.

21. Kamland Collaboration, "Partial Radiogenic Heat Model for Earth."

22. Frank P. Ryan, *The Mystery of Metamorphosis: A Scientific Detective Story* (White River Junction, Vt.: Chelsea Green, 2011); see also Ryan, "Metamorphosis: Nature's Most Transformative Process May Also Be an Unsung Force for Evolution," *New Scientist,* September 2011, 56–59.

23. Sonya E. Vickers and Donald I. Williamson, "Interspecies Hybrids," in Margulis, Asikainen, and Krumbein, *Chimeras and Consciousness,* 183–97; Donald I. Williamson and S. E. Vickers, "The Origin of Larvae," *American Scientist,* November–December 2007, 509–17.

24. The National Academy of Sciences rescinded this member privilege shortly thereafter. Members can no longer bypass editorial procedure to bring what they in their elected expertise think may be important overlooked contributions directly to their peers' attention.

12. METAMETAZOA

1. Empedocles, *Physics*, bk. 2, pt. 8.

2. Lynn Margulis, "Jointed Threads," *Natural History*, June 2005, 28–32.

3. *Science News*, March 30, 2002, http://www.sciencenews.org/view/access/id/24828.

4. For three domains, see, for example, "Towards a Natural System of Organisms: Proposal for the Domains Archaea, Bacteria, and Eukarya," *Proceedings of the National Academy of Sciences* 87 (1990): 4576–79. For Margulis's response to Woese's proposed phylogeny, see Lynn Margulis and Ricardo Guerrero, "Kingdoms in Turmoil," *New Scientist*, March 1991, 46–50. For five kingdoms together with three domains, see Lynn Margulis and Michael J. Chapman, *Kingdoms and Domains: An Illustrated Guide to the Phyla of Life on Earth* (New York: Elsevier/Academic Press, 2009). For a popular account, see William Brown, "A New Tree of Life Takes Root," *New Scientist*, August 11, 1990, 18.

5. The acceptance of symbiosis as a scientific fact of life has been championed in the previous century by Lynn Margulis: for details, see her *Symbiosis in Cell Evolution* (San Francisco: Freeman, 1982). A less technical narrative is presented in her *Early Life* (Boston: Jones and Bartlett, 1982). Margulis has demarginalized symbiosis theory, and the endosymbiotic origins of the eukaryotic cell are now presented as fact in many secondary- and college-level biology texts. Nonetheless, the theory of the origin of nucleated (eukaryotic) cells by symbiosis has been around for a century. For a history, see Jan Sapp, *Evolution by Association: A History of Symbiosis* (New York: Oxford University Press, 1994).

6. An ugly term, which sounds almost proctological, the term *protoctists* was coined in 1830 by the English naturalist John Hogg—the unenviability of whose own surname is suggested by the changing of the name of the Bahamas Hog Island to Paradise Island. While of little medical importance, protoctists are important both ecologically and evolutionarily: it is in this group of some thirty thousand species that plants, animals, and fungi evolved.

7. Michael W. Gray and W. Ford Doolittle, "Has the Endosymbiotic Theory Been Proven?" *Microbiological Reviews* 46 (1082): 1–42. See also John L. Hall and D. J. Luck, "Basal Body-Associated DNA: *In Situ* Studies in *Chlamydomonas reinhardtii*," *Proceedings of the National Academy of Sciences USA* 92 (1995): 5129–33; and, more recently, Hall, "Spirochete Contributions to the Eukaryotic Genome," *Symbiosis* 54 (2011): 119–29. As the symbiosis historian Jan Sapp underscores in "Freewheeling Centrioles," *History and Philosophy of the Life Sciences* 20 (1998): 255–90, one of the lessons of Hall's work is that cell structures do not have to have nucleic acids in order to be inherited. The study of nongenetic inheritance, epigenetics, has become a hot topic of active research in the years since I wrote this essay. "The classic example is the Dutch famine of World War II," writes Oded Rechavi. "Starving mothers who gave birth during the famine had children who were more susceptible to obesity and other meta-

bolic disorders—and so were their grandchildren." A study in rats showed that chronic high-fat diets in fathers result in obesity in their daughters. Rechavi is author of a paper showing that acquired traits can be inherited without DNA via small RNAs: Oded Rechavi, Gregory Minevich, and Oliver Hobert, "Transgenerational Inheritance of an Acquired Small RNA-Based Antiviral Response," in *C. elegans, Cell,* November 23, 2011, 1248–56.

8. Most evolutionary narratives are like mystery novels that leave out the beginning of the story. A popular account that does not make short shrift of the first three billion years of evolution is Lynn Margulis and Dorion Sagan, *Microcosmos: Four Billion Years of Microbial Evolution* (New York: Touchstone, 1991). See also A. Knoll, *Life on a Young Planet: The First Three Billion Years of Evolution on Earth* (Princeton, N.J.: Princeton University Press, 2004).

9. Best in the sense of ideal not most well. Unlike Hogg's less-than-mellifluous coinage, undulipodia seems an adequate, even euphonious word. From the Latin *undulatus,* "wavy," from the diminutive *undula,* "wavelet" "little wave," ultimately from *unda,* "wave," undulipodia refers to the waving appendages, be they cilia, sperm tails, or the cell whips of motile *Euglena* swimmers with the green solar eyes of their plastids. In an e-mail of January 19, 2012, F. J. R. "Max" Taylor, a Fellow of the Royal Society of Canada and an expert in symbiosis who coined the term *serial endosymbiosis theory,* told me, in the course of a discussion about how even he couldn't sign on to the spirochete idea, why he was reluctant to use the term *undulipodia*: "As I tried to make clear (and so did she) the symbiotic origin of mitochondria and chloroplasts proposals are old and gained acceptance as a result, not only of very strong structural and metabolic similarities with free living prokaryotes but, in particular, the fact that they are separated from the 'host' cytoplasm by their own membrane compartments and the presence of their own DNA inside them—I note that Lynn was one of the first to discover the latter in chloroplasts. Unfortunately, these are key features lacking in the basal bodies of flagella and cilia. Lynn would argue that this is because they became incorporated earlier but it would have helped her hypothesis if it was otherwise. Unfortunately I couldn't agree with her about restricting the term 'flagella' to bacteria since this is historically backwards—the name was not used for bacteria until nearly a century after it had been coined for eukaryotes (and I really disliked 'undulipodia')." It is perhaps ironic that the same general conservativism that leads life to preserve its form across the generations, and thus leave traces of its multibillion-year trajectory in the fossil record, also makes it, in the human linguistic realm, slow to adopt more accurate vocabulary terms. So it is that we say sunrise rather than, say, terraturn, or speak of dialing a phone when dials have long since been replaced by buttons and touch screens. The question is more than nomenclatural because of the size difference, distinctive ultrastructure, and the many more proteins found in the eukaryotic structures: some distinct word or sign, if not the six-syllable candidate in question, should be in use, if only for the sake of descriptive clarity among medical students and possible future researchers.

The plot thickens as Margulis also argued for a spirochete role in AIDS. Rather scandalously, soon after the announcement of the discovery of HIV as a cause of AIDS, reports of death by syphilis became conspicuously absent and was even remarked on by the British government. According to the independent researcher John Scythes, who has no institutional affiliations or axes to grind, but who has lost many friends to AIDs, syphilis remains a strong candidate for cofactor in the causation of symptoms attributed to HIV. A key datum is the immunocryptic abilities of syphilis spirochetes resistant to antibiotics. It may seem like a Lilliputian Freudian fantasy, but spirochetes really are the fastest beings in the microcosm, able to swim through viscous medium via a corkscrew movement that often lands them in the vicinity of food or other desirable gradients (light, heat, oxygen, anoxic environments) faster than their more sessile cousins; in termite hindguts they coordinate their undulations in waves; and they also not only feed on the periphery and inside the cell walls of "host" cells but also, remarkably, form permanent "holdfasts," permanently attaching to other cells, as in the notorious case, already mentioned, of *Mixotricha paradoxa,* a nanosphinx that has not only propulsive spirochetes attached to it but also congenital undulipodia, as well as at least two other permanently adopted bacterial genomes. With such witnessed behavior and morphologies, a minute sample of the vast earthly experiment of microbial evolution one must take seriously the proposal of an evolutionary spirochetosis at the base of the eukaryotic lineage.

As the theorized oldest partner in SET, serial endosymbiosis theory, anaerobic spirochetes may have merged with archaea millions of years before the addition of the ancestors to mitochondria or chloroplasts. And unlike those, theoretically more recent additions, the speedy spirochetes (their litheness and rapidity upping their chances for mergers) are both aerobic and anaerobic. Given the unsettled etiological and epistemological questions, and the ulterior motives of corporate profit, as well as the syphilis spirochete *Treponema pallidum*'s pre-AIDS notoriety as a or even "the great imitator," as chronicled in many old medical texts, that is a hardy slow killer able to simulate many other diseases (symptoms that may have largely overlapped with those of AIDS), the serious question arises as to whether, and if so, how, *T. pallidum,* known for its ability to go into hiding by forming immunologically indetectable "round bodies," is not itself (still) overwhelming human immune systems. To this end, as Scythes points out, it is good that syphilis, even if it is not admitted to be a possible cause or the cause of AIDS, is again becoming the subject of tests among AIDS patients. Considering that spirochetes have survived symbiotically as well as pathologically on Earth for well over three billion years, having survived all the major mass extinctions without missing a beat, it seems to me medically irresponsible not to carefully investigate, as Scythes advocates, their possible role as a cofactor of AIDS.

10. Georges Bataille's "general economy" and his constant reflections on the Sun are deeply influenced by Vladimir I. Vernadsky: see *Consumption,* volume 1 of *The Accursed Share:*

An Essay on General Economy (New York: Zone Books, 1988), especially 29 and 192, where Vernadsky is explicitly referred to. For Vernadsky in English, see his *The Biosphere* (Oracle, Ariz.: Synergetic, 1986), a much-abbreviated and perhaps unreliable "abridged version based on the French edition of 1929," and the more recent and comprehensive *The Biosphere: Complete Annotated Edition* (New Yorker: Springer, 1998). The new uses found for the excess materials produced in the wake of life's growth is a leitmotif of natural history. The wastes for which uses are found (e.g., oil deposits, calcium exudates, oxygen flatulence) produce, in turn, new wastes of their own. Pollution is not new, nor can it be attributed to the development of technology unless by technology we include nonhuman life-forms, among them bacteria. For further details on the uses to which wastes generated by rampant growths were put previously, see Margulis and Sagan, *Microcosmos,* pages 99–114 (for oxygen), 184–87 (for calcium), and 237 (for environmental crises in general).

11. For a meditation on the relation between spirochete microbial ecology and human thought, see Margulis's "Speculation on Speculation," in Margulis and Sagan, *Dazzle Gradually,* 48–56; and Margulis and Sagan, *Microcosmos,* 137–54.

12. See, for example, Dorion Sagan and Jessica Whiteside, "Medical Symbiotics," in Margulis, Asikainen, and Krumbein, *Chimeras and Consciousness,* 207–18. For new views of the immune system, see, for example, Gerard Eberl, "A New Vision of Immunity: Homeostasis of the Superorganism," *Mucosal Immunology* 3, no. 5 (2010): 450–60; and the review article by Yun Kyung Lee and Sarkis K. Mazmanian, "Has the Microbiota Played a Critical Role in the Evolution of the Adaptive Immune System?" *Science* 330 (2010): 1768–73.

13. Interview with Anthony Burgess, "Writers at Work," in *The Paris Review Interviews,* edited by George Plimpton, 4th ser. (New York: Penguin, 1976), 340–41.

14. The idea of a living Earth is not new: "Plato thought the world to be a living being and in *The Laws* (898) stated that the planets and stars were living as well. . . . During the Renaissance, the idea of Heaven as an animal reappeared in Lucillo Vanini; the Neoplatonist Marsilio Ficino spoke of the hair, teeth and bones of the earth; and Giordano Bruno felt that the planets were great peaceful animals, warm-blooded, with regular habits and endowed with reason. At the beginning of the seventeenth century, the German astronomer Johannes Kepler debated with the English mystic Robert Fludd which of them had first conceived the notion of the earth as a living monster, 'whose whale like breathing, changing with sleep and wakefulenss, produces the ebb and flow of the sea.' The anatomy, the feeding habits, the colour, the memory and the imaginative and shaping faculties of the monster were sedulously studied by Kepler" (Jorge Luis Borges, *The Book of Imaginary Beings* [New York: Penguin, 1980], 21–22).

15. For Gaia theory as unscientific, see, for example, Heinrich Holland, *The Chemical Evolution of the Atmosphere and Oceans* (Princeton, N.J.: Princeton University Press, 1984); W. Ford Doolittle, "Is Nature Really Motherly?" *CoEvolution Quarterly* 29 (1981): 58–65; and

R. Dawkins, *The Extended Phenotype* (Oxford: Freeman, 1982). For Gaia theory as trivial, see J. W. Kirchner, "The Gaia Hypothesis: Can It Be Tested?" *Review of Geophysics* 27 (1989): 223–35. For "Satanism," see Carol White, "Mother Earth Marries Satan," *Twenty-First Century Science and Technology*, September–October 1989, 52–53.

16. Cited in James E. Lovelock, *The Ages of Gaia: A Biography of Our Living Earth* (New York: Norton, 1988). The astronaut gazes at the baby-blue, cloud-flecked planet from which she or he is now separated. Earth, spoken of anemically in textbooks as lifeless, a mere geochemical setting for life, no longer appears as mere environment. It mutates from being the home of an ecology or ecofeminism and becomes a giant spherical being. For astronaut accounts, see Frank White, *The Overview Effect: Space Exploration and Human Evolution* (Boston: Houghton Mifflin, 1987).

17. Although truth may be stranger than fiction, fiction is often truer—if only because its claims to represent truth are less strident. Mythopoetic realities are freely generated within the realm of science fiction. A living planet thrives in Isaac Asimov's book *Foundation and Earth*. In the Polish writer Stanisław Lem's *Solaris*, a planet is inhabited by a giant ocean capable of copying human artifacts and even human beings. In R. A. Kennedy's novel *The Trinuniverse*, Mars divides into something like a giant cell and begins feeding on other planets. In *Born of the Sun*, the science fiction writer Jack Williamson portrays the planets of our solar system as ova laid by the Sun, Earth being the first to hatch. In my book *Biospheres* (see next note), a putative work of science nonfiction, I extended the Gaian metaphor of aliveness to the point of reproduction. In this logical extension of the Gaian trope of a live Earth, I pictured the surface planetary environment as a neuter being (rather than a "goddess") on the verge of potentially stellar reproduction (but not necessarily self-conscious of that fact). Such stellar reproduction borders on the incredible and, partly because of that, illustrates a noble lie of the Gaian kind.

18. Dorion Sagan, *Biospheres: The Metamorphosis of Planet Earth* (New York: Bantam/McGraw-Hill, 1990).

19. The group (Space Biospheres Ventures) initially in charge of the Biosphere 2 project was a scientifically pathetic "cult" on a "bogus journey." See Marc Cooper, "Take This Terrarium and Shove It," *Village Voice*, April 2, 1991, 24–33, and "Profits of Doom: The Biosphere Project Finally Comes out of the Closet—as a Theme Park," *Village Voice*, July 30, 1991, 31–36.

The biggest problem, apart from leaching of atmospheric oxygen from inside the structure by its cement (making inhabitant "biospherians" so weak they could not ascend to the library [where there was a copy of my book *Biospheres*]), was that the company took pains to conceal the fact that the structure was not completely sealed and even required an electric generator in its basement. Now, however, Biosphere 2 is a scientific research facility owned by the University of Arizona.

20. Bataille, *Accursed Share*, 29.

21. Sorin Sonea and Maurice Panisset, *The New Bacteriology* (Boston: Jones and Bartlett, 1983).

22. See chapter 1 for the startling findings of the biologist Kwang Jeon, who witnessed the transformation of infectious bacteria into the needed organelles of a new species of amoeba. Jeon may be the only person in human history to have actually witnessed, at least at such closeness and so single-handedly, the evolution of a new species in the laboratory.

23. See, for example, Mariel Emrich, "Cancer Cell Mitochondria," *Science Friday, Talking Science*, http://www.talkingscience.org/2012/01/cancer-cell-mitochondria/.

24. Sorin Sonea and Leo G. Mathieu, *Prokaryotology: A Coherent View* (Montreal: Les Presses de L'Université de Montréal).

25. The Homeric epics never mention a body—the flesh-enclosed entity usually taken for granted as the definable material self—but speak only of what we would think of as body parts, corporeal fragments such as "fleet legs" and "sinewy arms"; see Bruno Snell, *The Discovery of the Mind*, translated by Thomas E. Rosenmeyer (New York: Harper Torchbooks, 1960), 8. "The idea of the self in a case," Norbert Elias has written, "is one of the recurrent *leitmotifs* of a modern philosophy, from the thinking subject of Descartes, Leibniz' windowless monads and the Kantian subject of knowledge (who from his aprioristic shell can never quite break through to the 'thing in itself') to the more recent extension of the same basic idea of the entirely self-sufficient individual" (*The Civilizing Process: The History of Manners*, trans. E. Jephcott [New York: Urizen, 1978], 252–53).

26. James E. Lovelock, *Gaia: A New Look at Life on Earth* (New York: Oxford University Press, 1979).

27. James E. Lovelock, "Life Span of the Biosphere," *Nature*, April 1982, 561–63.

13. KERMITRONICS

1. Erwin Schrödinger, *What Is Life? The Physical Aspect of the Living Cell* (Cambridge: Cambridge University Press, 1944).

2. Guiseppe Trautteur, "The Illusion of Free Will and Its Acceptance," in *After Cognitivism: A Reassessment of Cognitive Science and Philosophy*, edited by Karl Leidlmair (New York: Springer, 2009), 192–203.

3. For full Einstein poem, facsimile, and discussion, see http://www.physicsforums.com/showthread.php?t=127828.

4. Masao Matsuhashi and Mark Hallett, "The Timing of the Conscious Intention to Move," *European Journal of Neuroscience* 28 (2008): 2344–51.

5. Heinrich Von Kleist, "On the Theater of Marionettes," in *Selected Prose of Heinrich von*

Kleist (Brooklyn: Archipelago Books, 2009), 265–74; online translation by Idris Parry, http://southerncrossreview.org/9/kleist.htm.

14. ON DOYLE ON DRUGS

1. Cited in Richard Doyle, *Darwin's Pharmacy: Sex, Plants, and the Evolution of the Noösphere* (Seattle: University of Washington Press, 2011), 74.

2. Lynn Sagan, "Communications: An Open Letter to Mr. Joe K. Adams," *Psychedelic Review* 1, no. 3 (1964): 355.

3. Ibid., 354.

4. Doyle, *Darwin's Pharmacy,* 73.

5. Gerald Schueler, "Chaos and the Psychological Symbolism of the Tarot," http://www.schuelers.com/chaos/chaos7.htm.

6. Ironically perhaps, this paean to the miraculous nature of books was not written but televised, in episode 11, "The Persistence of Memory," of the *Cosmos* series.

7. The noösphere refers to the "thought-sphere" of humans and technology around the Earth. The term is associated with the Catholic evolutionist Pierre Teilhard de Chardin, but both he and *Édouard Le Roy* attended lectures at the Sorbonne by Vernadsky, who used it more secularly. The geologist Eduard Suess in 1875 coined the term *biosphere,* placing life in the context of other material spherical systems at Earth's surface: the atmosphere, the hydrosphere, the geosphere. Vernadsky and Chardin met, but the latter was more focused on seeing a Singularity-like teleology in evolution's future, a final state when man could achieve his goal and blend into spirit (again): the so-called Omega Point. Vernadsky, more subtly I think, looked at it as an extension of the linked mineral and thermodynamic transformations worked by life with the Sun's energy. Before Earth was seen from space, Vernadsky recognized life as a global phenomenon. In fact, he popularized the term *biosphere.* Whether the human transport of munitions or biblically reported locust clouds transmuting fields of grain into flying mountains, life was a geological force. To the mineralogists and crystallographer Vernadsky, global industry and telecommunications was a material extension of the biosphere.

8. Jacques Derrida, "Plato's Pharmacy," in *Dissemination,* translated by Barbara Johnson (Chicago: University of Chicago Press, 1981).

9. Terrence McKenna, *Food of the Gods: A Radical History of Plants, Drugs, and Human Evolution* (New York: Bantam, 1993).

10. E. O. Wilson, *The Social Conquest of Earth* (Brooklyn: Liveright, 2012).

11. Jacques Derrida, *Paper Machine,* translated by Rachel Bowlby (Palo Alto, Calif.: Stanford University Press, 2005), 6.

12. Jacques Derrida, *Of Grammatology,* translated by Gayatri Chakravorty Spivak (Baltimore, Md.: Johns Hopkins University Press, 1976).

13. Eric D. Schneider and D. Sagan, *Into the Cool: Energy Flow, Thermodynamics, and Life* (Chicago: University of Chicago Press, 2005).

14. Geoffrey Miller, *The Mating Mind: How Sexual Choice Shaped the Evolution of Human Nature* (New York: Anchor, 2001).

15. Jeffrey Wicken, *Evolution, Thermodynamics, and Information: Extending the Darwinian Program* (New York: Oxford University Press, 1987).

CONCLUSION

1. Sharon E. Kingsland, "The Beauty of the World: Evelyn Hutchinson's Vision of Science," in *The Art of Ecology: Writings of G. Evelyn Hutchinson*, edited by David K. Skelly, David M. Post, and Melinda D. Smith (New Haven, Conn.: Yale University Press, 2010).

2. E. D. Schneider and J. J. Kay, "Life as a Manifestation of the Second Law of Thermodynamics," *Mathematical Computer Modeling* 19 (1994): 25–48.

3. Aldous Huxley, *The Perennial Philosophy* (London: Chatto and Windus, 1946).

4. J. Scott Turner, *The Extended Organism: The Physiology of Animal-Built Structures* (Cambridge, Mass.: Harvard University Press, 2000); see also Turner, "Gaia, Extended Organisms, and Emergent Homeostasis," in *Scientists Debate Gaia: The Next Century,* edited by Stephen H. Schneider, James R. Miller, Eileen Crist, and Penelope J. Boston (Cambridge, Mass: MIT Press, 2004), 57–70.

5. E-mail message to author, May 5, 2012.

6. Stephen Jay Gould, *Rocks of Ages: Science and Religion in the Fullness of Life* (New York: Ballantine Books, 1999), 3.

7. Heinz Von Foerster, "On Constructing a Reality," in *Environmental Design Research,* vol. 2, edited by Wolfgang F. E. Preiser (Stroudsburg, Penn.: Dowden, Hutchinson, and Ross, 1973), 35–46. As with Twain's comment on the exaggerated news of his own death, we should not be too quick to pronounce benedictions over Cartesianism's body. A fascinating, still relevant book, advocating self-improvement through reprogramming one's own mind, is Maxwell Maltz, *Psycho-Cybernetics: A New Way to Get More Living Out of Life* (Upper Saddle River, N.J.: Prentice-Hall, 1960).

8. Stefan Helmreich, "*Homo microbis* and the Figure of the Literal," Culture@Large 2011, "Theorizing the Contemporary," Society for Cultural Anthropology, 2012, http://www.culanth.org/?q=node/511 and http://vimeo.com/37271969.

9. Sigmund Freud, *Beyond the Pleasure Principle* (New York: Norton, 1989), 45–46.

10. Jacques Derrida, *Archive Fever: A Freudian Impression*, translated by Eric Prenowitz (Chicago: University of Chicago Press, 1998), 19–20.

11. Ibid., 35.

12. Isaiah 45:17, quoted in Pir Zia Inayat-Kahn, "A Hidden Treasure," http://www.sevenpillarshouse.org/article/a_hidden_treasure (accessed May 20, 2012).

13. Derrida, *Archive Fever,* 2.

14. Kim TallBear, "Why Interspecies Thinking Needs Indigenous Standpoints" (2012), http://www.culanth.org/?q=node/510; and vimeo, http://www.culanth.org/?q=node/509. Link also includes responses by Myra Hird and Augustin Fuentes to essay 1.

15. Bruce Scofield, "Gaia: The Living Earth. 2500 Years of Precedents in Natural Science and Philosophy," in *Scientists Debate Gaia: The Next Century,* edited by Stephen H. Schneider, James R. Miller, Eileen Crist, and Penelope J. Boston (Cambridge, Mass.: MIT Press, 2004), 151–60.

INDEX

Dorion Sagan is an award-winning science writer, editor, and theorist. He has written or coauthored more than two dozen books on culture, evolution, and the history and philosophy of science, including *What Is Life?*, *Into the Cool*, and *Death and Sex*. His writing has been published in the *New York Times*, the *New York Times Book Review*, *Wired*, *Natural History*, *Times Higher Education*, *Smithsonian*, and *Cabinet*. He is the son of the astronomer Carl Sagan and the biologist Lynn Margulis.